NEW METHODS OF CONCURRENT CHECKING

FRONTIERS IN ELECTRONIC TESTING

Consulting Editor
Vishwani D. Agrawal

Books in the series:

New Methods of Concurrent Checking
Goessel, M., Ocheretny, V., (et al.), Vol. 42
ISBN 978-1-4020-8419-5
Digital Timing Measurements – From Scopes and Probes to Timing and Jitter
Maichen, W., Vol. 33
ISBN 0-387-32418-0
Fault-Tolerance Techniques for SRAM-based FPGAs
Kastensmidt, F.L., Carro, L. (et al.), Vol. 32
ISBN 0-387-31068-1
Data Mining and Diagnosing IC Fails
Huisman, L.M., Vol. 31
ISBN 0-387-24993-1
Fault Diagnosis of Analog Integrated Circuits
Kabisatpathy, P., Barua, A. (et al.), Vol. 30
ISBN 0-387-25742-X
Introduction to Advanced System-on-Chip Test Design and Optimi...
Larsson, E., Vol. 29
ISBN: 1-4020-3207-2
Embedded Processor-Based Self-Test
Gizopoulos, D. (et al.), Vol. 28
ISBN: 1-4020-2785-0
Advances in Electronic Testing
Gizopoulos, D. (et al.), Vol. 27
ISBN: 0-387-29408-2
Testing Static Random Access Memories
Hamdioui, S., Vol. 26
ISBN: 1-4020-7752-1
Verification by Error Modeling
Redecka, K. and Zilic, Vol. 25
ISBN: 1-4020-7652-5
Elements of STIL: Principles and Applications of IEEE Std. 1450
Maston, G., Taylor, T. (et al.), Vol. 24
ISBN: 1-4020-7637-1
Fault injection Techniques and Tools for Embedded systems Reliability…
Benso, A., Prinetto, P. (Eds.), Vol. 23
ISBN: 1-4020-7589-8
Power-Constrained Testing of VLSI Circuits
Nicolici, N., Al-Hashimi, B.M., Vol. 22B
ISBN: 1-4020-7235-X
High Performance Memory Memory Testing
Adams, R. Dean, Vol. 22A
ISBN: 1-4020-7255-4
SOC (System-on-a-Chip) Testing for Plug and Play Test Automation
Chakrabarty, K. (Ed.), Vol. 21
ISBN: 1-4020-7205-8
Test Resource Partitioning for System-on-a-Chip
Chakrabarty, K., Iyengar & Chandra (et al.), Vol. 20
ISBN: 1-4020-7119-1
A Designers' Guide to Built-in Self-Test
Stroud, C., Vol. 19
ISBN: 1-4020-7050-0

NEW METHODS OF CONCURRENT CHECKING

by

Michael Goessel
Potsdam University,
Germany

Vitaly Ocheretny
Infineon Technologies AG,
Neubiberg, Germany

Egor Sogomonyan
Potsdam University,
Germany

and

Daniel Marienfeld
Potsdam University,
Germany

Springer

Prof. Dr. Michael Gössel
Universität Potsdam
Inst. Informatik
14439 Potsdam
Germany
mgoessel@cs.uni-potsdam.de

Dr. Vitaly Ocheretny
Infineon Technologies AG
Am Campeon 1-12
85579 Neubiberg
Germany
vitalij@cs.uni-postdam.de

Prof. Dr. Egor Sogomonyan
Universität Potsdam
Inst. Informatik
14439 Potsdam
Germany

Dr. Daniel Marienfeld
Universität Potsdam
Inst. Informatik
14439 Potsdam
Germany

ISBN: 978-90-481-7876-6 e-ISBN: 978-1-4020-8420-1

Printed on acid-free paper

9 8 7 6 5 4 3 2 1

springer.com

Preface

Computers are everywhere around us. We, for example, as air passengers, car drivers, laptop users with Internet connection, cell phone owners, hospital patients, inhabitants in the vicinity of a nuclear power station, students in a digital library or customers in a supermarket are dependent on their correct operation.

Computers are incredibly fast, inexpensive and equipped with almost unimaginable large storage capacity. Up to 100 million transistors per chip are quite common today - a single transistor for each citizen of a large capital city in the world can be easily accommodated on an ordinary chip. The size of such a chip is less than 1 cm^2.

This is a fantastic achievement for an unbelievably low price. However, the very small and rapidly decreasing dimensions of the transistors and their connections over the years are also the reason for growing problems with reliability that will dramatically increase for the nano-technologies in the near future.

Can we always trust computers? Are computers always reliable? Are chips sufficiently tested with respect to all possible permanent faults if we buy them at a low price or have errors due to undetected permanent faults to be discovered by concurrent checking? Besides permanent faults, many temporary or transient faults are also to be expected.

What do we know about the detection of these transient faults during normal operation?

Contrary to permanent faults, transient faults cannot be detected by testing but rather only by concurrent checking, and the design of effective error detection circuits for concurrent checking will be a challenging problem in the next years.

The book in front of you, "New Methods of Concurrent Checking", answers the question as to how the best possible state-of-the-art error detection circuits can be designed.

This book describes the latest new and effective methods for concurrent checking for digital circuits which were developed mainly in the last 15 years. Some of the methods are published for the first time.

The book is of interest to students and teachers in electrical engineering and computer science, researchers and designers and all readers who are interested in error detection and reliability of digital circuits and computers.

The authors have worked together in research and industrial projects in the area of concurrent checking in the Fault-Tolerant Computing Group at the University of Potsdam, Germany. This group was founded in 1992 as a Max-Planck-Working Group and was incorporated into the Institute of Computer Science of the university in 1997.

Many of the ideas and results presented here were obtained in cooperation with guest researchers from different parts of the world. We thank all of them.

We are especially grateful to our colleagues V. Saposhnikov and Vl. Saposhnikov at the University of Transportation, St. Petersburg, for their many wonderful ideas and contributions, especially in the areas of self-dual error detection and error detection by complementary circuits. The basic ideas in these areas arose in close cooperation with them as can also be seen from the corresponding references.

Personally we would also like to thank K. Chakrabarty, Duke University, for his stimulating discussions and advice over the years, and also the former Ph.D. students A. Morosov, A. Dmitriev, A. Morosov, M. Seuring, M. Moshanin and H. Hartje for their valuable contributions.

We are grateful to Paul Synnott for proofreading this book.

The authors

Contents

1 **Introduction** 1

2 **Physical Faults and Functional Errors** 5
2.1 Stuck-At Faults 6
2.2 Bridging Faults 10
2.2.1 Non-Resistive Bridging Faults 10
2.2.2 Resistive Bridging Faults 13
2.3 CMOS Stuck-Open and Stuck-On Faults 17
2.4 Delay Faults 21
2.5 Transient Faults 22
2.6 Functional Error Model 22
2.7 Output Dependencies 24
2.8 Self-Testing and Self-Checking 25
2.9 Faults and Errors in Submicron Technologies 27

3 **Principles of Concurrent Checking** 31
3.1 Duplication and Comparison 31
3.1.1 Description of the Method 32
3.1.2 Comparators and Two-Rail Checkers 33
3.1.3 Method of Partial Duplication 40
3.2 Block Codes for Error Detection 42
3.2.1 Classical Error Detection Codes 42
3.2.2 Non-linear Split Error Detection Codes 49
3.3 Parity and Group Parity Checking 53
3.3.1 Predictor and Generator Circuits 55
3.3.2 Parity Prediction 57
3.3.3 Generalized Circuit Graph 60
3.3.4 Independent Outputs and Weakly Independent Outputs 62
3.3.5 Determination of Groups of Weakly Independent Outputs 66
3.3.6 Circuit Modification by Node-Splitting 70
3.3.7 Further Methods for the Determination of Weakly Independent Outputs 73

3.4 Odd and Even Error Detection ... 75
3.4.1 Description of Odd and Even Error Detection ... 75
3.5 Code-Disjoint Circuits ... 77
3.5.1 Design of Code-Disjoint Circuits ... 78
3.6 Error Detection by Complementary Circuits ... 84
3.6.1 Error Detection by Use of Complementary Circuits ... 85
3.6.2 Complementary Circuits for 1-out-of-3 Codes ... 87
3.6.3 Conditions for the Existence of Totally Self-Checking Error Detection Circuits by Complementary Circuits ... 90
3.7 General Method for the Design of Error Detection Circuits ... 98
3.7.1 Description of the Method ... 98
3.8 Self-Dual Error Detection ... 102
3.8.1 Self-Dual Boolean Functions ... 103
3.8.2 Transformation of a Given Circuit into a Self-Dual Circuit ... 104
3.8.3 Self-Dual Error Detection Circuits ... 107
3.8.4 Self-Dual Fault-Secure Circuits ... 109
3.9 Error Detection with Soft Error Correction ... 116
3.9.1 Description of the Method ... 116

4 Concurrent Checking for the Adders ... 123
4.1 Basic Types of Adders ... 124
4.2 Parity Checking for Adders ... 135
4.3 Self-Checking Adders ... 138
4.3.1 Self-Checking Carry Look-Ahead Adders ... 138
4.3.2 Self-Checking Partially Duplicated Carry Skip Adder ... 151
4.3.3 Self-Checking Carry Select Adders ... 156

References ... 173

Index ... 179

Chapter 1
Introduction

No technical system is 100% reliable. This is true for modern chips and will be much more relevant for the next generation of chips.

The following types of errors may be distinguished:

1. Specification errors
2. Design errors
3. Production errors
4. Errors due to deliberate attacks.

Only the detection of errors caused by physical faults will be considered in this book although, in principle, also errors evoked by deliberate attacks can be detected by similar methods.

The percentage of undetected erroneously functioning devices used can be reduced by *testing*, by *concurrent checking* or *on-line detection* and by *fault tolerance*.

The notions of testing, concurrent checking and fault tolerance are explained in brief below.

- *Testing*
  After production every chip is tested. In a *test* a predetermined set of input values, a test set, is applied to the chip, which is called the *Circuit Under Test* (CUT). If, for the inputs from that test set, the outputs of the chip are not as expected, the CUT is erroneous. If these outputs are correct, the CUT is considered to be fault-free.
  In an external test the test inputs from the test set are supplied by an external tester. In *Built-In Self-Test* (BIST) the test inputs are generated on the chip. Different methods of testing are described, for instance, in [1, 2, 3].
  Most, but not all, of the permanent faults can be detected by testing. The percentage of faulty chips which are not detected by the test depends on the fault coverage of the test and the percentage of faulty chips before testing.
- *Concurrent Checking*
  Temporary or transient faults are active only for a short duration of time, and the moments of their occurrences are not predictable. Therefore, transient faults

cannot be detected in advance by testing. The errors caused by transient faults have to be detected during normal operation.
Error detection during normal operation is called *concurrent checking* or *on-line detection*.
In concurrent checking or on-line detection the outputs of the considered functional circuit are constantly monitored by use of redundant hardware. Only the functional inputs, and no specially selected inputs, are applied during normal operation.
The simplest and best known method of concurrent checking is duplication and comparison.
The monitored circuit C is duplicated in a functionally equivalent circuit C'. The functional inputs are applied in parallel to both the circuits C and C' and the outputs of C and C' are compared by a comparator. An erroneous output of one of the circuits C or C' is detected by the comparator provided the comparator is fault-free.
Since the necessary area overhead and the power consumption for duplication and comparison are high, other methods of concurrent checking have been developed.

- *Fault Tolerance*
In a *fault-tolerant system* errors are compensated to some extent by the system. The best-known method for the design of a fault tolerant system is triple modular redundancy (TMR).
The considered system S is triplicated in three functionally identical systems $S_1 = S_2 = S_3 = S$.
The same functional inputs are applied in parallel to all three systems S_1, S_2 and S_3, and a majority voter V determines from the outputs of S_1, S_2 and S_3 the output of the triplicated system.
If one of the systems S_1, S_2 or S_3 fails, the output of one of the systems may be erroneous and (at least) two will be correct. The majority voter V determines from the two correct outputs and the single erroneous output the correct output as the output of the triplicated system. A fault within one of the systems S_1, S_2 or S_3 will be corrected and the triplicated system is fault tolerant with respect to such a fault.
Also error correction codes, (for instance Hamming codes,) error detection and reconfiguration and other methods are used in fault-tolerant designs. Fault-tolerant systems are described, for instance, in [4, 5, 6].

"New Methods for Concurrent Checking" considers concurrent checking or on-line detection for digital circuits.

New methods of concurrent checking are presented, most of which were developed within the last 15 years.

Duplication and comparison are, as already pointed out, conceptually the simplest method of error detection.

No special assumptions for the expected errors or faults are needed as long as the faults or errors are located within one of the duplicated circuits.

All methods of error detection other than duplication and comparison have to be compared to this method with respect to the error detection probability, the necessary area and the power consumption.

In practice, a method of concurrent checking is of interest if the necessary area is considerably smaller than the 220–250% of the area of the functional circuit needed for duplication and comparison, and if the probability of detecting errors due to single stuck-at faults is about $90\%+x$.

The contents of the following chapters of the book are presented in brief below.

Physical faults and how these faults cause functional errors are described in Chapter 2. Stuck-at faults, bridging faults, transistor stuck-open and stuck-on faults, delay faults, the functional error model corresponding to a set of physical faults and different output dependencies as independent outputs, weakly independent outputs and unidirectionally independent outputs are explained. These dependencies between circuit outputs are used in the Chapters 3 and 4 for the design of error detection circuits.

Chapter 3 contains basic and many new principles of concurrent checking for random logic. The chapter starts with duplication and comparison, partial duplication with and without parity checking for the non-duplicated circuit part. Attention is given to self-checking and easily testable comparators. A new easily testable comparator is presented for the first time.

Error detection using systematic codes is considered in detail. A new class of non-linear codes for error detection is introduced. Besides errors which are detected with certainty almost all the remaining errors are detected with a probability greater than $1/2$ by these non-linear codes.

The generalized circuit graph shown in [7] is described. This graph expresses the connections of the gates of the considered circuit in summary. The nodes of this graph are the maximal classes of gates with one output. The generalized circuit graph is utilized to determine groups of independent and weakly independent outputs.

It is shown how these groups of independent and weakly independent outputs of a given combinational circuit can be utilized for an optimal design of error detection circuits by means of parity and group parity codes.

To improve the error detection capability of parity and group parity codes, circuit transformations – described as "node splitting" of the gates belonging to fan-out nodes of the generalized circuit graph - are considered.

It is shown how an arbitrarily given parity-checked circuit can be easily transformed into a code-disjoint circuit with one or two additional outputs. Serial connections of code-disjoint circuits are also considered.

A completely new method of error detection by complementary circuits is presented. For a given functional circuit with n outputs, a complementary circuit, also with n outputs, is added such that the componentwise *XOR* sum of the corresponding outputs of the functional circuit and of the complementary circuit is an element of a chosen code. Also *m-out-of-n* codes and not only systematic codes can be used.

It is shown that the design space for the complementary circuit is huge and some heuristic design methods are described.

For a given combinational circuit, conditions necessary and sufficient for the existence of a totally self-checking circuit by use of a complementary circuit are proven for an *1-out-of-n* code.

It is shown how the design of an error detection circuit with a specified functional error model can be reduced to a standard synthesis problem, the optimum design of a partially defined circuit. Because of the limited capabilities of the synthesis tools for circuit optimization, this general method can in practice only be applied to small circuits, and the different structural methods for the design of optimum error detection circuits remain of great importance.

Self-dual error detection, including self-dual parity and self-dual duplication, is explained. For this method the necessary area overhead is low but the original input and the corresponding inverted input must always be successively applied to the inputs of the self-dual circuit. Therefore, the time is increased by a factor of two and this method is applicable to systems in which time is not critical. Such systems are, for instance, control systems of mechanical systems.

At the end of this chapter the combination of the correction of soft errors in the register cells by use of C-elements and error detection by systematic codes is considered.

Chapter 4 is concerned with error detection for regular structures. The different types of adders are chosen as regular structures. The methods developed in Chapter 3 for random logic can be more specifically applied and slightly modified to obtain optimum results for other regular structures like adders. A fast carry ripple adder, different carry look-ahead adders, carry skip adders and carry select adders are considered.

A new *sum-bit-duplicated* adder cell is introduced and used for concurrent checking complementary to the well-known carry-duplicated adder cell. Based on the concept of error detection by partial duplication, the sum-bit-duplicated adder cell is utilized for different code-disjoint adder designs. Since the output registers of the corresponding adders are also duplicated, all soft errors in these output registers, odd and even, are detected by comparing the contents of these duplicated registers. The non-duplicated part of the adder is checked by parity prediction. Without duplicating the complete adder, the error detection probability of this sum-bit-duplicated adder is almost the same as for the duplicated adder.

For other regular structures, such as different types of multipliers or dividers, similar results can be obtained – as recently published in the literature [8, 9, 10].

It is not the only intention of Chapter 4 for the different adder types to develop the best possible error detection circuit but also to demonstrate how the general methods of error detection can be applied to regular structures and to encourage the designers of future chips to develop their own optimum self-checking designs in the technology available.

The contents of the book were used in the "Fault Tolerant System Design" lectures in the last 7 years for undergraduate and graduate students at the University of Potsdam, Germany and in some tutorials by one of the authors.

We hope the reader will be inspired by the ideas presented in this book to work in this interesting field and will have the same pleasure the authors have over the past years.

Chapter 2
Physical Faults and Functional Errors

Combinational and sequential circuits are built up of gates and memory elements. Gates and memory elements for their part are currently made of transistors. Gates and memory elements may be faulty due to physical faults.

Examples of physical faults are a short to ground or a shortage in the power supply of a line, bridging between lines, broken lines, faults caused by α-particles and high-energy neutrons, faults due to electromagnetic fields, including crosstalk. Faults may be permanent or transient. Transient faults are active only for a short time.

In this chapter different types of physical faults and functional errors will be considered. An explanation is given as to how different faults can be modeled and how faults cause functional errors.

A description is given of how single stuck-at faults occur if a short to ground or a shortage in the power supply of a circuit line occurs. It will be demonstrated how a single stuck-at fault causes an error at a circuit output if an error is stimulated at the location of the fault and if the stimulated error is propagated to a circuit output along a sensitized path.

It will be shown that the probability of propagating an error to a circuit output decreases exponentially with the distance of the location of the fault from the circuit output. The distance is measured by the number of gates with a controlling value. Based on this result it will also be shown that the probability of a two-bit error is significantly lower than the probability of a single-bit error.

It is then demonstrated that bridging faults occur if two circuit lines are erroneously connected. Non-resistive and resistive bridging faults are considered. In this chapter we assume that non-resistive bridging faults may be adequately modeled by an *OR*-function.

For resistive bridging faults it is explained how an unknown continuous parameter, the resistance R of the bridge, determines the different types of possible errors.

The functional effects of bridging faults differ according to circumstances, and simple examples are used to illustrate that a bridging fault can be modeled either as a single stuck-at fault, as a replacement of an *AND*-gate by an *OR*-gate, by an abstract automaton or by an oscillating signal line.

Stuck-open and stuck-on *CMOS* transistor faults will be considered in detail, and it will be demonstrated that these transistor faults may result in a sequential behavior of the circuit or in an undefined value with a high leakage current.

Gate delay and path delay faults will be distinguished.

It will be discussed under what conditions single-event transitions in the combinational part of the circuit generate soft errors in the memory elements, and that soft errors directly induced in the memory elements are much more frequent.

Physical faults can erroneously alter the functional behavior of a circuit.

It will be shown how a functional error model can be determined. An error function or an error automaton will be assigned to every physical fault of the considered fault model. The error model for single gates gives examples of single stuck-at faults and bridging faults.

Output dependencies on faults will be introduced as useful error models. Independent outputs, weakly independent outputs and unidirectionally independent outputs will be discussed. These dependencies express how faults may simultaneously influence pairs or groups of circuit outputs. In the following chapters these dependencies will be utilized for optimal design of error detection circuits.

To quantify the quality of error detection by concurrent checking the notions of self-testing, fault-secure and total self-checking for circuits with concurrent checking are explained.

To also guarantee the detection of input errors code-disjoint circuits will be considered.

All these notions will be defined with respect to an arbitrary fault model. Usually these notions apply for the fault model of all single stuck-at faults of the considered circuit.

It will be discussed in detail that, although at present in reality many types of faults other than single stuck-at faults occur, these definitions with respect to single stuck-at faults remain of real value.

The established belief is that the detection of all errors caused by all single stuck-at faults will also result in a satisfactory error detection probability for errors due to other types of faults. The similarity of this concept with the idea of an n-detection test for single stuck-at faults to also detect other faults will be discussed.

At the end of this chapter the main reasons why concurrent checking is becoming increasingly important for submicron technologies will be given.

2.1 Stuck-At Faults

The single stuck-at fault model is the most frequently used fault model, and almost all error detection circuits are designed using this model. A circuit line is said to be stuck-at 0 (1) if the line is fixed to 0 (1).

The single stuck-at model will be explained for an *OR*-gate, which is shown in Fig. 2.1

The input lines of the *OR*-gate in Fig. 2.1 are labeled by 1 and 2 and the output line by 3 respectively.

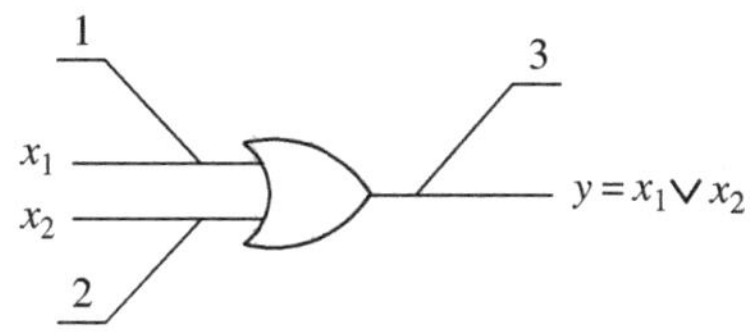

Fig. 2.1 *OR*-gate

If the line 1 is stuck-at i, this stuck-at fault will be denoted by l/i. If the *OR*-gate is fault-free, it implements the Boolean function

$$y = x_1 \vee x_2.$$

Now we assume that line 1 is stuck at 0, which will be described as $1/0$. Then the faulty *OR*-gate implements

$$y_{1/0} = 0 \vee x_2 = x_2$$

instead of $y = x_1 \vee x_2$.

For all the single stuck-at faults $1/1$, $2/1$ and $3/1$ the faulty *OR*-gate implements

$$y_{1/1} = y_{2/1} = y_{3/1} = 1.$$

The truth table for the correct *OR*-gate and for the faulty *OR*-gates with the single stuck-at faults $1/0$ and $1/1$ are given in Table 2.1.

It can be seen from Table 2.1 that the single stuck-at fault $1/0$ causes an error at the output of the *OR*-gate for the input combination $x_1 = 1$, $x_2 = 0$ and that for the single stuck-at fault $1/1$ the output of the *OR*-gate is erroneous for the inputs $x_1 = 0$, $x_2 = 0$.

To understand under what conditions internal stuck-at faults of a combinational circuit cause errors at the circuit outputs, the notion *controlling value* and *non-controlling value* of a gate are useful.

An input value v, $v \in \{0, 1\}$ of a gate G is a controlling value if the output value of G is uniquely determined by v. If v is a controlling value, then the output of G is independent of the input value at the remaining input of G.

If v is not a controlling value, then v is called a non-controlling value.

For an *AND*-gate, $v = 0$ is a controlling value since for $x_1 = v = 0$ the output $y = x_1 \wedge x_2 = 0 \wedge x_2 = 0$ is independent of the value of x_2. $v = 1$ is a non-controlling

Table 2.1 Truth table of an *OR*-gate with single stuck-at faults 1/0 and 1/1

x_1	x_2	y	$y_{1/0}$	$y_{1/1}$
0	0	0	0	1
0	1	1	1	1
1	0	1	0	1
1	1	1	1	1

value. For $x_1 = v = 1$ the output $y = x_1 \wedge x_2 = 1 \wedge x_2$ depends on the values of the second input x_2.

For *AND*, *NAND*, *OR* and *NOR*-gates the controlling values are 0, 0, 1 and 0 and the non-controlling values are 1, 1, 0 and 0 respectively.

For *XOR* and *XNOR*-gates only non-controlling values exist. For both the input values 0 and 1 at the first input of these gates the output depends on the value of the second input.

To better understand how an error at a circuit output is generated by an internal single stuck-at fault we consider the circuit of Fig. 2.2

The circuit has two outputs y_1 and y_2 and nine inputs $x_1, x_2, \ldots, x_9$. The eight gates are numbered from 1 to 8. (For simplicity of presentation, the signals $x_1, x_2, \ldots, x_9$ are considered as external inputs. They may also be internal signals of a larger circuit.)

We assume that there is a single stuck-at fault $1/1$ at the input line 1 of gate G_1.

An error at the input line 1 of gate G_1 which is caused by the single stuck-at fault 1/1 is stimulated under the input $x_1 = 0$. The stimulated error is propagated to the first circuit output y_1 along a first path consisting of the gates G_1, G_2, G_3 and G_4 if all the side inputs x_2, x_3, x_4, x_5 of the gates G_1, G_2, G_3 and G_4 of the path from the location of the fault $1/1$ to the circuit output y_1 are non-controlling values. This is the case for $x_2 = 1$, $x_3 = 0$, and $x_4 = 1$. Since the gate G_4 is an *XOR*-gate, every input of this gate is a non-controlling value and the side input x_5 is arbitrary.

An error is stimulated and propagated to the circuit output y_1 only for the value assignment $x_1 = 0, x_2 = 1, x_3 = 0$ and $x_4 = 1$. If we assume that the values for all the input variables are randomly distributed equally, then the probability that an error results at the output y_1 due to the considered stuck-at fault is $1/16$.

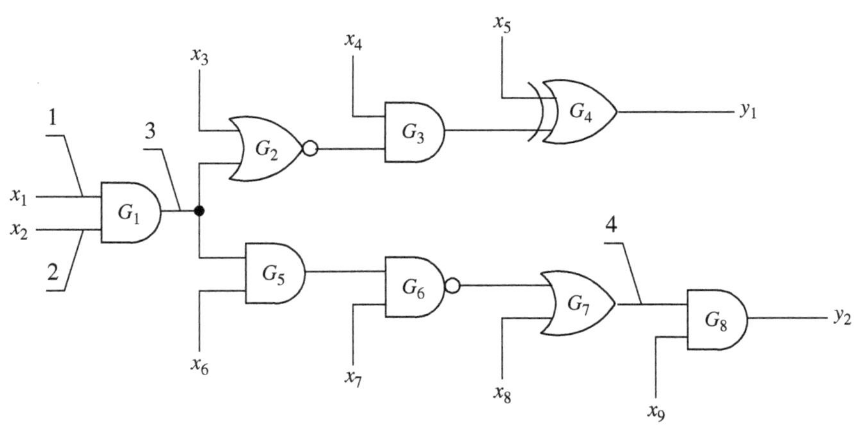

Fig. 2.2 An example of a combinational circuit

Similarly, the single stuck-at fault $1/1$ causes an error at the second output y_2 if the error is stimulated by $x_1 = 0$ and propagated along the sensitized second path consisting of the gates G_1, G_5, G_6, G_7 and G_8. Again, the side inputs of the gates of this path have to be the non-controlling values $x_2 = 1, x_6 = 1, x_7 = 1, x_8 = 0$ and $x_9 = 1$. If we again assume that the values for the input variables are randomly and equally distributed, the probability that the single stuck-at fault $1/1$ causes an error at the circuit output y_2 is $1/64$.

A single-bit error occurs if an error is stimulated and propagated through gate G_1 and propagated along only one of the paths G_2, G_3, G_4 or G_5, G_6, G_7, G_8, but not on both.

The probability of stimulating and propagating the error through gate G_1 is $\frac{1}{4}$. Along the first path G_2, G_3, G_4, where G_4 is an *XOR*-gate, the error will be propagated with a probability of $\frac{1}{4}$ and along the second path G_5, G_6, G_7, G_8 with a probability of $1/16$. Thus, the probability p_{1bit} of a single-bit error is

$$p_{1bit} = \frac{1}{4} \times (\frac{1}{4} + \frac{1}{16} - \frac{1}{4} \times \frac{1}{16}) = \frac{19}{256} \approx 0.074$$

Both the circuit outputs y_1 and y_2 are simultaneously erroneous if the error is stimulated by $x_1 = 0$ and simultaneously propagated along the paths G_1, G_2, G_3, G_4 and G_1, G_5, G_6, G_7 and G_8. This is only possible if the side inputs of both of these paths are the non-controlling values $x_2 = 1, x_3 = 0, x_4 = 1$ and $x_2 = 1, x_6 = 1, x_7 = 1, x_8 = 0, x_9 = 1$.

Since a two-bit error occurs only for a single value assignment of the eight variables $x_1, x_2, x_3, x_4, x_6, x_7, x_8, x_9$ out of $2^8 = 256$ possible values, the probability p_{2bit} that a two-bit error is caused by the single stuck-at-1 fault $1/1$ is $1/256$.

$$p_{2bit} = \frac{1}{4} \times \frac{1}{16} = \frac{1}{256} \approx 0.004.$$

As illustrated by the sample circuit of Fig. 2.2 a stimulated error will be propagated along a path of gates to a circuit output if all the side inputs of the gates are non-controlling values. Since all input values of *XOR*-gates and *XNOR*-gates are non-controlling values, an error is always propagated through *XOR* and *XNOR*-gates.

Let us consider a path of N_{contr} gates with a controlling value and N_{ncontr} gates with no controlling value.

If we assume that the values for the side inputs of the gates on the path are randomly equally distributed with probabilities of the values 1 and 0 both equal to $1/2$, then the probability $p_{prop}(e)$ that a stimulated error e will be propagated to a circuit output along the considered path is

$$p_{prop}(e) = 2^{-N_{contr}}. \tag{2.1}$$

According to equation (2.1) the probability that a stimulated error is propagated from the location of the fault to a circuit output exponentially decreases with the *distance* of the location of the fault from the circuit output. More precisely, this distance is determined by the number of gates with a controlling value.

If there are now two paths from the location of a fault to two circuit outputs y_1 and y_2 of $N_1 = N_{1,contr} + N_{1,ncontr}$ and $N_2 = N_{2,contr} + N_{2,ncontr}$ gates, where $N_{i,contr}$ and $N_{i,ncontr}$ for $i = 1,2$ are the number of gates with and without controlling values respectively.

Then the probability $p_{prop}^{two}(e)$ that a fault will be simultaneously propagated from the location of the fault to the two circuit outputs y_1 and y_2 is

$$p_{prop}^{two}(e) = 2^{-(N_{1,contr}+N_{2,contr})}, \tag{2.2}$$

and the probability that a stimulated error is propagated simultaneously to two circuit outputs y_1, y_2 exponentially decreases with the sum of the distances of the outputs y_1 and y_2 from the location of the fault.

Compared to the probability $p_{prop}^{single}(e)$

$$p_{prop}^{single}(e) = 2^{-N_{1,contr}} + 2^{-N_{2,contr}} - 2^{-(N_{1,contr}+N_{,contr})} \tag{2.3}$$

that the stimulated error e is propagated exactly to a single one of these outputs, the probability $p_{prop}^{two}(e)$ of a two-bit error is low.

2.2 Bridging Faults

A bridging fault occurs if two or more wires are connected due to a physical fault.

2.2.1 Non-Resistive Bridging Faults

In the simplest case of a non-resistive bridging fault an additional (small) resistance between the erroneously connected wires is neglected. Depending on the technology, a *wired OR* or a *wired AND* may result. More specifically, bridging faults may be also considered at the transistor level [3]. For simplicity of presentation we assume in this chapter that a bridging fault is functionally described by a *wired OR*. If a bridging fault results in a *wired AND*, the considerations are similar.

Let us consider an *AND*-gate represented in Fig. 2.3 with the input lines 1 and 2 and the output line 3.

First we assume that there is a bridging fault $brid(1,2)$ between the input lines 1 and 2. Then the erroneous *AND*-gate is to be functionally replaced by the circuitry of Fig. 2.4

The bridging between the input lines is functionally described by an *OR*-gate. The circuit of Fig. 2.4 implements the erroneous output y_e with

$$y_e = (x_1 \vee x_2) \wedge (x_1 \vee x_2) = x_1 \vee x_2.$$

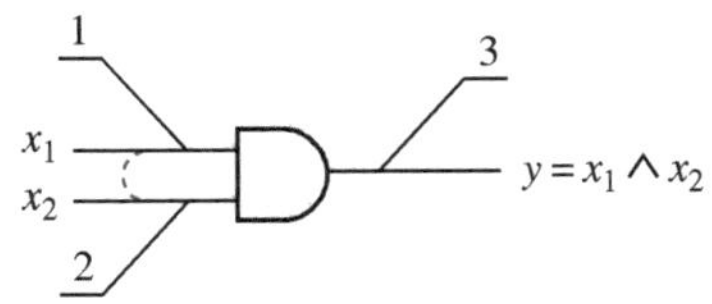

Fig. 2.3 *AND*-gate with bridging between the input lines 1 and 2

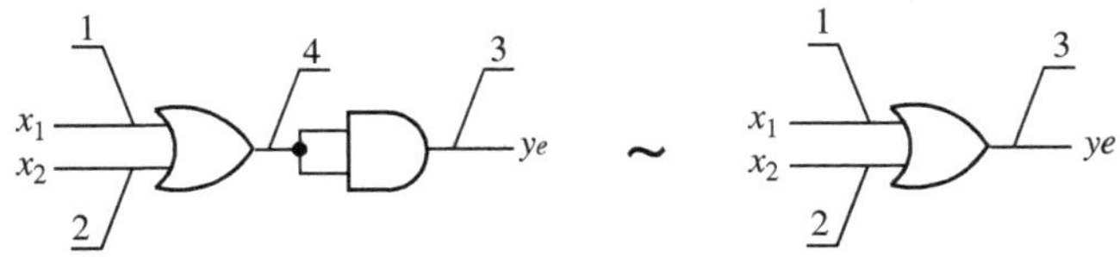

Fig. 2.4 Functional model of input bridging of Fig. 2.3

The *AND*-gate with a bridging fault at its input lines is equivalent to an *OR*-gate.

As a further example we consider an input-output bridging fault of an *OR*-gate as shown in Fig. 2.5.

Due to this bridging fault the erroneous *OR*-gate becomes a simple state machine. Depending on the input $x_1(t)$, $x_2(t)$ and the output $y_e(t)$ at time t the output $y_e(t+1)$ at time $t+1$ is determined as shown in Table 2.2.

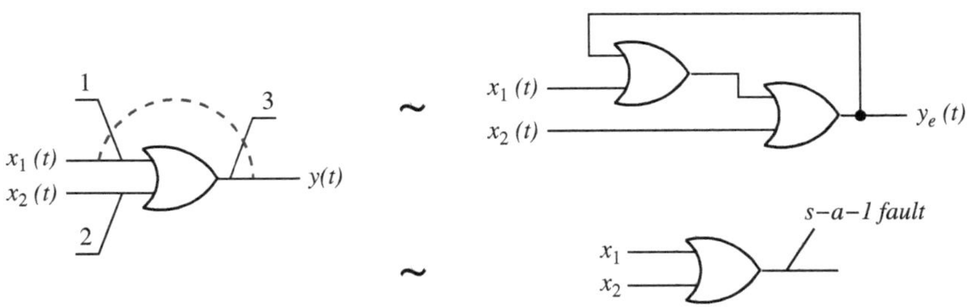

Fig. 2.5 Input-output bridging of an *OR*-gate and functionally equivalent circuits

Table 2.2 Input-output bridging fault of an *OR*-gate

$x_1(t)$	$x_2(t)$	$y_e(t)$	$y_e(t+1)$
0	0	0	0
0	0	1	1
0	1	0	1
0	1	1	1
1	0	0	1
1	0	1	1
1	1	0	1
1	1	1	1

If $y_e(t) = 0$, then for the inputs $x_1(t) = x_2(t) = 0$ the output $y_e(t+1)$ remains 0. If $y_e(t) = 1$, the output $y_e(t+1)$ remains 1 for the same inputs $x_1(t) = x_2(t) = 0$, and the circuit that functionally describes an input-output bridging fault of a combinational *AND*-gate is sequential.

But, as can be seen from Table 2.2, if the output is $y_e(t) = 1$ for the first time, then the output will be equal to 1 "forever" and the bridging fault in Fig. 2.5 can in reality be described as a single stuck-at-1 fault of the output of the considered *OR*-gate.

Figure 2.6 illustrates an input-output bridging fault $brid(1,3)$ between the input line 1 and the output line 3 of the *AND*-gate of Fig. 2.3.

The behavior of the *AND*-gate with this bridging fault is described by Table 2.3 and by the state diagram of an automaton with two states $y_e = 0$ and $y_e = 1$ in Fig. 2.7.

If we apply the sequence (0,0); (0,1); (1,0); (1,1) to the erroneous *AND*-gate of Fig. 2.6, the output sequence according to Table 2.3 or the state diagram of Fig. 2.7 is 0; 0; 0; 1 and no error is indicated. But, if we apply the same four inputs in a different order - (1,1); (0,1); (1,0); (0,0) - the corresponding output sequence will be 1, 1, 0, 0 and in the second position an error occurs.

Even if this *AND*-gate with the considered bridging fault is "exhaustively" tested with all its four possible $2^2 = 4$ input values in the order (0,0); (0,1); (1,0); (1,1), the bridging fault $brid(1,3)$ will not be detected. But if during normal operation the input (1,1) is followed by the input (0,1), an erroneous output value 1 instead of the correct output value 0 is caused by this bridging fault.

A completely different situation may occur if in the feedback path of input-output bridging an inverter is included as is shown for a *NAND*-gate in Fig. 2.8.

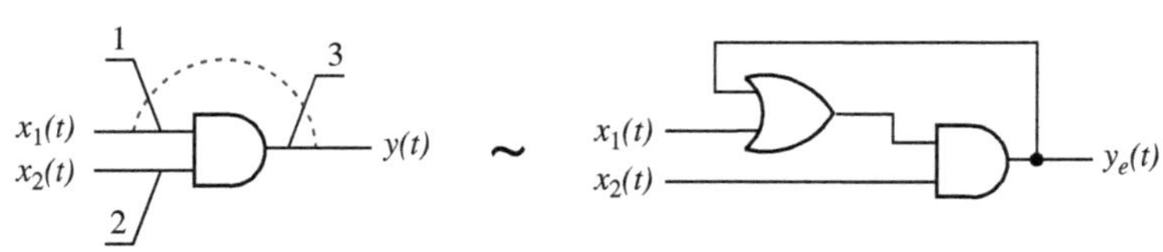

Fig. 2.6 Input-output bridging of an *AND*-gate and functionally equivalent circuit

Table 2.3 Sequential circuit describing an input-output bridging fault of an *AND*-gate

$x_1(t)$	$x_2(t)$	$y_e(t)$	$y_e(t+1)$	y_{corr}
0	0	0	0	0
0	0	1	0	0
0	1	0	0	0
0	1	1	1	0
1	0	0	0	0
1	0	1	0	0
1	1	0	1	1
1	1	1	1	1

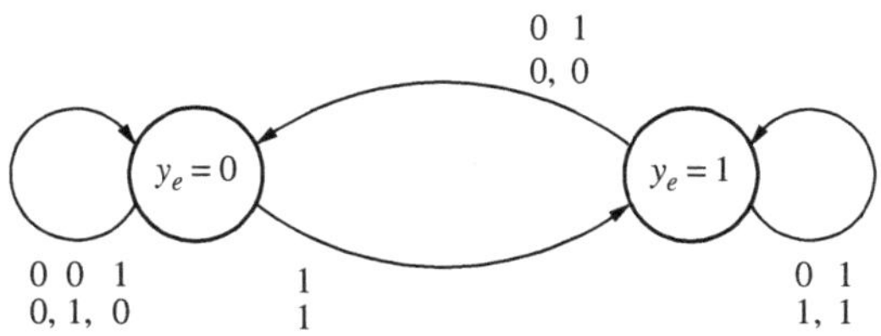

Fig. 2.7 State diagram of the automaton corresponding to input-output bridging of an *AND*-gate

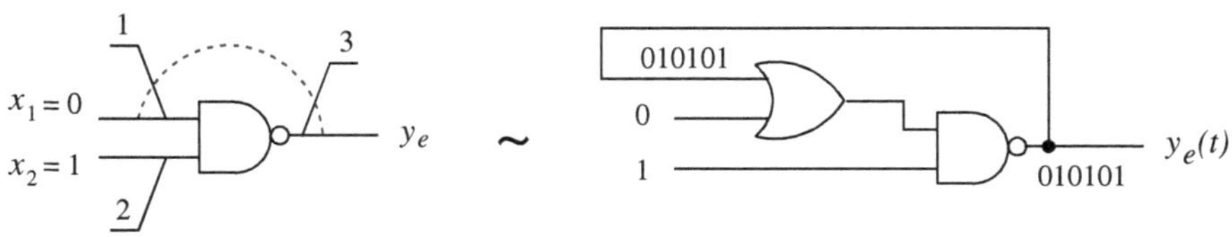

Fig. 2.8 Input-output bridging with an inverter in the feedback path

For the input values $x_1 = 0$, $x_2 = 1$ the erroneous output y_e is oscillating. No value, neither 0 nor 1, at the output y_e of the erroneous *NAND*-gate is stable.

The number of possible bridging faults is huge. If the number of gates of the considered circuit is *N*, then the number of circuit lines is proportional to *N* and the number of bridging faults between two lines is proportional to N^2. For a relatively moderate number $N = 10^7$ of gates there are about 10^{14} bridging faults to be considered.

If layout information is available, only bridging faults between circuit lines which are close in the layout are to be considered. These are the bridging faults between the inputs and the outputs of the $N = 10^7$ gates and their neighboring gates in the layout.

In reality it can rarely be assumed that all bridging faults, including the input-output bridging faults, which result in a sequential behavior of the circuit have been detected in the test mode after production. Therefore, especially in the modern nano-technologies, undetected bridging faults may be expected which may result in erroneous outputs during normal operation. The errors caused by the bridging faults which were not detected by testing have to be detected by concurrent checking.

For a more realistic description, bridging faults have to be modeled as resistive bridging faults.

2.2.2 Resistive Bridging Faults

A short between circuit lines, which is called a bridging fault, is modeled as a *resistive bridging fault*, if the electrical resistance between the erroneously connected wires is taken into account. If a resistive bridging fault occurs between internal gate lines [11] the fault is regarded as internal. If the short takes place between input and output lines of gates, the bridging fault is regarded as external.

Figure 2.9 shows the model of a concrete external resistive bridging fault, a short between the wires W_1 and W_2, the output lines of the predecessor gates G_1 and G_2 of the bridge. G_1 is an inverter and G_2 is a *NAND*-gate. The wires W_1 and W_2 are supposed to be, due to a physical fault, electrically connected via an electrical resistance R.

The main problem of resistive bridging faults is that *the values of the resistances of the possibly bridged pairs of wires are continuous unknown parameters of the model.*

In Fig. 2.9 the wires W_1 and W_2 are input lines of the successor gates G_3 and G_4, where G_3 is an *OR*-gate and G_4 a *NAND*-gate. The side inputs of these gates are x_4 and x_5 respectively.

The resistive bridging fault between the wires W_1 and W_2 results in a functional error if, due this fault, for some input x_1, x_2, x_3, x_4, x_5 at least one of the outputs y_1 or y_2 of the successor gates G_3 of G_4 is erroneous.

For a large resistance, ideally for $R = \infty$, no fault occurs. For very small values of R, ideally for $R = 0$, the bridging fault is non-resistive.

If both the values v_1 on W_1 and v_2 on W_2 are equal, i.e. for

$$v_1 = \bar{x}_1 = v_2 = \overline{x_2 \wedge x_3},$$

the bridging fault has no logical effect. If the values v_1 on W_1 and v_2 on W_2 are different, i.e. for

$$v_1 = \bar{x}_1 \neq v_2 = \overline{x_2 \wedge x_3},$$

the resistive bridging fault induces instead of the correct values v_1 and v_2 some intermediate voltage levels u_1 and u_2 with $0 < u_1, u_2 < 1$ on the wires W_1 and W_2.

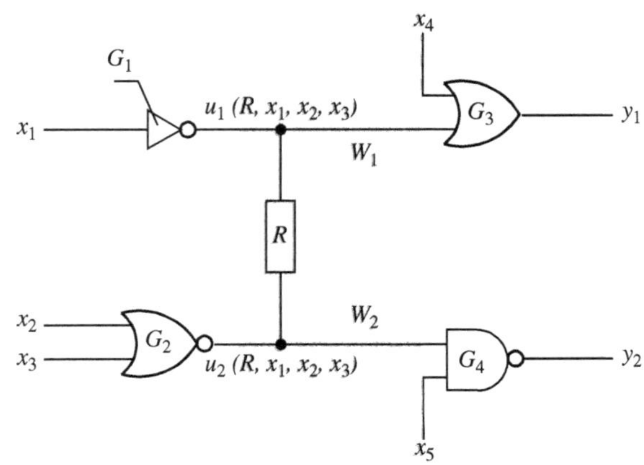

Fig. 2.9 Example of an external resistive bridging fault

If the intermediate voltage level u_1 is larger than or equal to the threshold value thr_3 of gate G_3, the intermediate voltage level u_1 will be interpreted by the successor gate G_3 as 1 (high), otherwise as 0 (low).

Similarly, if the intermediate voltage level u_2 is larger than or equal to the threshold value thr_4 of the gate G_4, u_2 will be interpreted by the successor gate G_4 as 1, otherwise as 0.

A resistive bridging fault with the resistance R can be detected by a test T if the value of the resistance R is within a specific interval which is called the "Analog Detection Interval" [12].

More precisely, if, in the presence of the resistive bridging fault, for $R \in [0, R_{max}]$ at least for one test vector of the test set T an erroneous output is generated, the interval $[0, R_{max}]$ is called an *Analog Detection Interval ADI* of the fault with respect to the considered test set T [12, 13]. (Also more complicated cases with non-continuous Analog Detection Intervals are possible.)

R_{max} is the maximum value of the resistance for which the considered bridging fault is detectable by a test vector from the given test set T.

The intermediate voltage levels u_1 and u_2 depend on the resistance R of the bridge and on the number of conducting transistors of the predecessor gates, which is determined by the actual values of the corresponding inputs of these gates. Also the technological parameters of the transistors are of influence.

In the considered example the predecessor gates are G_1 and G_2, and the conducting transistors are determined by the concrete input values of x_1, x_2 and x_3.

We illustrate this for the input $x_1 = 0$ of G_1 and for the two different pairs of inputs $x_2 = 1, x_3 = 0$ and $x_2' = 1, x_3' = 1$ of G_2.

A CMOS-implementation of the *NOR*-gate G_2 is shown in Fig. 2.10.

For $x_1 = 0$ the correct output of the inverter G_1 is

$$v_1 = \overline{x}_1 = \overline{0} = 1.$$

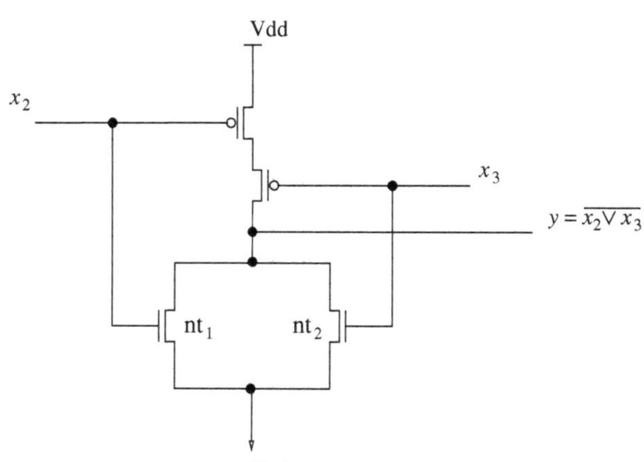

Fig. 2.10 *CMOS* implementation of *NOR*-gate G_2

Both the correct outputs $v_2 = \overline{x_2 \vee x_3}$ and $v_2' = \overline{x_2' \vee x_3'}$ are equal to zero

$$v_2 = \overline{1 \vee 0} = 0$$

and

$$v_2' = \overline{1 \vee 1} = 0.$$

For $x_2 = 1, x_3 = 0$ and $x_2' = 1, x_3' = 1$ the output y of the *NOR*-gate is disconnected from VDD. For $x_2 = 1$ and $x_3 = 0$ the output y is connected to GND only via the n-transistor nt_1 and for $x_2' = 1$, $x_3' = 1$ via both the n-transistors nt_1 and nt_2.

Without a resistive bridging fault the output y of gate G_2 is the same for $x_2 = 1$, $x_3 = 0$ and for $x_2' = 1$, $x_3' = 1$, but, in the case of the considered resistive bridging fault, the intermediate voltage levels u_1, u_2 and u_1', u_2' are different.

The logical values y_3 of gate G_3 and y_4 of gate G_4 depend on the intermediate voltage levels and on their thresholds values thr_3 and thr_4 respectively. The threshold values thr_3 and thr_4 depend also on the side inputs x_4 and x_5 of G_3 and G_4 respectively.

For $x_1 = 0$ the correct value on the wire W_1 is 1. In the presence of the considered resistive bridging fault we have for the intermediate voltage level $u_1 < 1$.

Let $x_4 = 0$ be the non-controlling value of the *OR*-gate G_3. For

$$u_1 \leq thr_3(x_4)$$

the output y_3 is erroneously changed from 1 to 0, and for

$$u_1 > thr_3(x_4)$$

the output $y_3 = 1$ remains correct. If $x_4 = 1$ is the controlling value of gate G_3, the output $y_3 = 1$ of G_3 remains always correct.

For $x_2 = 1$, $x_3 = 0$ or for $x_2' = 1$, $x_3' = 1$ the correct value on the wire W_2 is 0. In the presence of the considered resistive bridging fault, we have $u_2 > 0$ for the intermediate voltage level u_2.

If $x_5 = 1$ is the non-controlling values of gate G_4 then for

$$u_2 \leq thr_4(x_5)$$

the output y_2 is erroneously changed from 1 to 0, and for

$$u_2 > thr_4(x_5)$$

the output $y_2 = 1$ remains correct. If $x_5 = 0$ is equal to the controlling value of the *NAND*-gate G_4, the output $y_2 = 1$ is determined by this controlling value and is always correct.

Depending on the resistance R of the bridging fault, the inputs x_1, x_2, x_3, x_4, x_5 and the threshold values thr_3 and thr_4 the following cases are possible:

1. y_1 is correct, y_2 is correct,
2. y_1 is erroneous, y_2 is correct,

3. y_1 is correct, y_2 is erroneous,
4. y_1 is erroneous, y_2 is erroneous.

In the fourth case an internal 2-bit error occurs.

Depending on the concrete circumstances and the concrete successor gates of a resistive bridge a two-bit error may occur which can be either unidirectional (0,0 into 1,1 or 1,1 into 0,0) or bidirectional (1,0 into 0,1 or 0,1 into 1,0).

This may be a rare situation, and in most cases at least one of these internal two-bit errors will not be propagated to a circuit output and simultaneously latched in a register, but the described situation requires special attention for concurrent checking with respect to internal two-bit errors.

In principle, resistive bridging faults, with the exception of some feedback bridging faults, can be detected to some extent by testing. For the unknown parameter R a probability distribution $p(R)$ is either determined from experimental data of the chips which are to be tested or by some reasonable assumptions and an expected fault detection probability can be computed [12].

Concurrent checking is of special importance for the detection of errors due to feedback bridging faults.

It was shown in [14], [11] by simulation that, depending on the value of the resistance R, a ciruit line may be oscillating between intermediate voltage levels with a frequency much higher than the clock frequency of the circuit. The detection of such an error during testing depends on the concrete timings of the strobes of the tester and cannot be guaranteed.

Bridging faults with high resistances will not have an immediate logical effect but they may result in small timing variations [11]. For the high clock frequencies of modern computers this type of error is also of growing importance and has to be detected by concurrent checking.

2.3 CMOS Stuck-Open and Stuck-On Faults

Typical transistor faults in *CMOS* technology are now considered. In *CMOS* technology, gates and circuits are implemented as a p-net and a complementary n-net.

A typical *CMOS* implementation of a combinational function $f(x)$ is shown in Fig. 2.11.

The p-net implements the function $f(x)$ by p-transistors, and the n-net the inverted function $\overline{f}(x)$ by n-transistors. In the fault-free case, either the p-net or the n-net is conductive depending on the input x. If, for the input x the p-net is conductive, then VDD (~ 1) is connected to the output and $y(x) = 1$. The output is connected to GND (~ 0) if the n-net is conductive. Then we have $y(x) = 0$.

Since in the fault-free case the p-net and the n-net are at no time simultaneously conductive VDD is never connected to GND and, besides the switching activity, there is (in an ideal case) no current flowing between VDD and GND.

If we now assume that either the p-net or the n-net is faulty, then the following situations are possible:

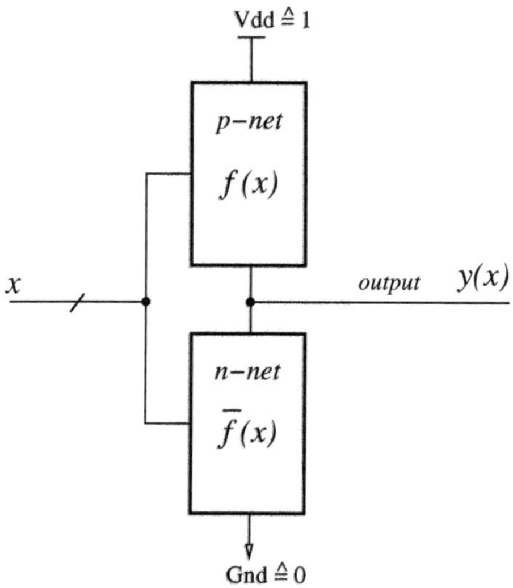

Fig. 2.11 *CMOS* implementation of a boolean function $f(x)$

1. For an input x both the p-net and the n-net are conductive.
2. For an input x both the p-net and the n-net are simultaneously non-conductive.

In the first case, if the p-net and the n-net are simultaneously conductive, the output $y(x)$ is connected both to VDD (~ 1) and to GND (~ 0) and the actual output value is undefined, $y(x) = u$.

Depending on the actual value of u, which is mainly determined by the resistances of the p-net and n-net, u may be interpreted as 1 or 0.

In the second case both the p-net and the n-net are not conducting and the output is disconnected from both VDD (~ 1) and GND (~ 0). Then the actual output value $y(x(t))$ remains the previous output value $y(x(t-1))$.

The circuit implementing a combinational function in the fault-free case becomes sequential due to this fault.

Now some simple examples of actual transistor faults will be considered.

A transistor is a voltage-controlled resistance [15]. For the simplicity of presentation, the p- and n-transistors are modeled as switches.

P- and n-transistors are shown in Figs. 2.12 and 2.13.

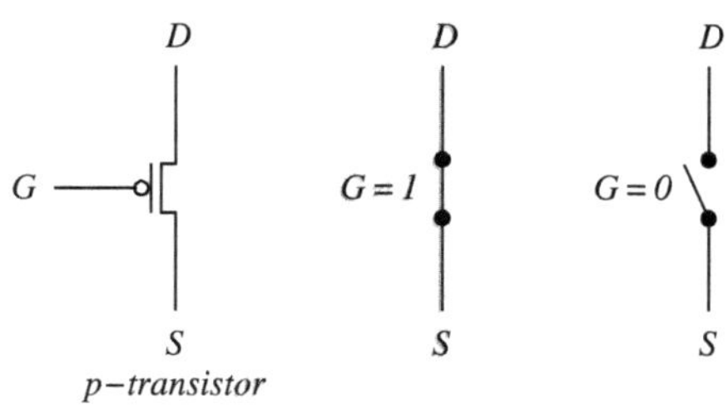

Fig. 2.12 Switching model of a p-transistor

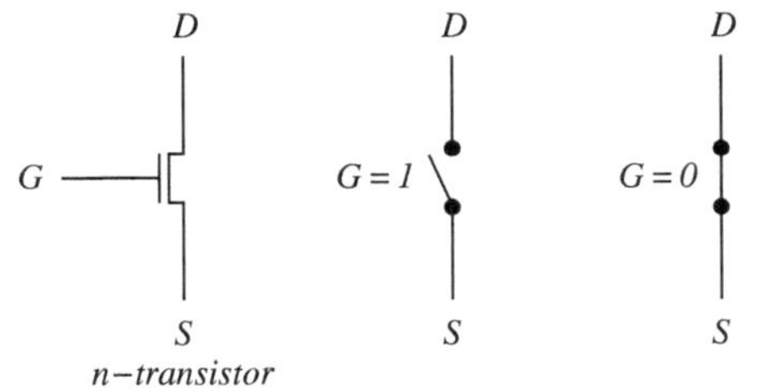

Fig. 2.13 Switching model of an n-transistor

For a p-transistor in Fig. 2.12, D (drain) is connected to S (source) if $G(gate) = 1$, and D is not connected to S if $G = 0$.

An n-transistor is shown in Fig. 2.13. For an n-transistor, D is connected to S if $G = 0$, and D is not connected to S if $G = 1$.

Stuck-open and stuck-on faults are now explained:

- *Stuck-open fault of a transistor*:
  If D (drain) and S (source) are, independent of the value of G (gate), always disconnected, the transistor is regarded as being stuck open.
  A stuck-open fault can be modeled as an open switch.
- *Stuck-on fault of a transistor*:
  If D (drain) and S (source) are, independent of the value of G (gate), always connected, the transistor is regarded as being stuck on.
  A stuck-on fault can be modeled as a closed switch.

To explain the logic errors caused by transistor faults we now consider a *CMOS* implementation of a *NOR*-gate, which is presented in Fig. 2.14.

The p-net consists of the p-transistors A and B and the complementary n-net of the n-transistors C and D. For $x_1 = x_2 = 0$ the output y is connected to VDD (~ 1). In all other cases, y is connected to GND (~ 0).

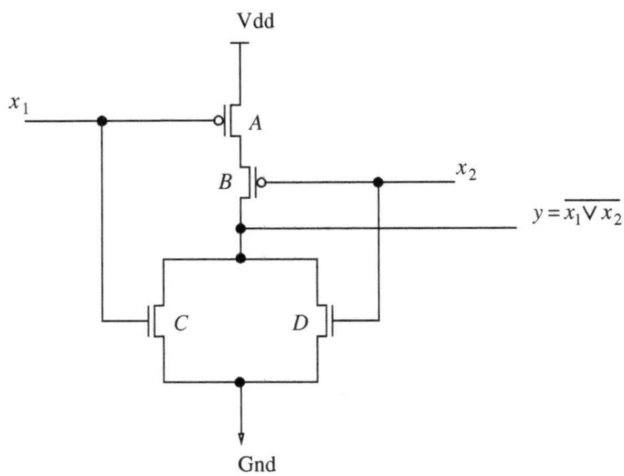

Fig. 2.14 *CMOS* implementation of a *NOR*-gate

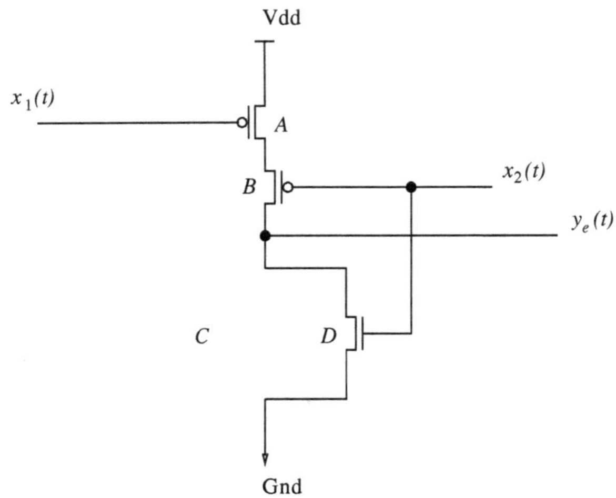

Fig. 2.15 *CMOS NOR*-gate with a stuck-open fault of a transistor *C*

Let us assume now that the transistor *A* is stuck open. Then *VDD* (~ 1) is always disconnected from the output y. For the input $x_1 = x_2 = 0$ *GND* (~ 0) is also disconnected from y and the previous output value will be retained. In principle this will be either 0 or 1. But from the time the output of the faulty *NOR*-gate is 0 for the first time it will remain 0 forever and the stuck-open fault of the transistor *A* can be sufficiently well modeled by a single stuck-at 0 fault of the output y of the *NOR*-gate.

Next we consider a stuck-open fault of the transistor *C*. In this case there is no connection from *GND* via the transistor *C* to the output y. The corresponding faulty *NOR*-gate is shown in Fig. 2.15. The behavior of the faulty *NOR*-gate of Fig. 2.15 is determined by Table 2.4.

For $x_1(t) = 1$, $x_2(t) = 0$ the output $y_e(t)$ is disconnected from both *VDD* and *GND* and the previous output $y_e(t-1)$ is retained. The faulty *NOR*-gate becomes sequential.

Table 2.4 Table of values for a *NOR*-gate with a stuck-open transistor

$x_1(t)$	$x_2(t)$	$y_e(t)$	$y_e(t+1)$
0	0	0	1
0	0	1	1
0	1	0	0
0	1	1	0
1	0	0	0
1	0	1	1
1	1	0	0
1	1	1	0

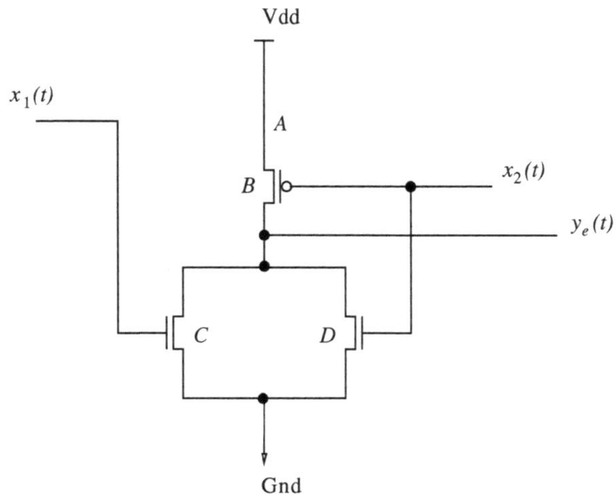

Fig. 2.16 *CMOS NOR*-gate with a stuck-on fault of a transistor *A*

If this faulty *NOR*-gate is "exhaustively" tested with all possible four inputs ordered as 11, 10, 01, 00, the corresponding output sequence will be 0, 0, 0, 1 and all the outputs are correct.

But if the input sequence is applied in the order 00, 10, 01, 11, the corresponding output sequence of the faulty *NOR*-gate is 1, 1, 0, 0 and the fault will be detected.

A stuck-open transistor fault may not be detected even if all possible input combinations are applied to the inputs of the faulty gate but not in the necessary order.

This is similar to an input-output bridging fault with no inverter in the feedback line.

Let the transistor *A* now be stuck on. Then the transistor *A* is always conducting and the corresponding faulty *NOR*-gate is shown in Fig. 2.16.

For $x_1 = 1$ and $x_2 = 0$ the output y_e is simultaneously connected to both *VDD* (~ 1) and *GND* (~ 0) and the value of the output y_e is undefined u, $y = u$. It cannot be predicted whether y_e will be interpreted as 1 or 0 and this fault may not be detected by a logic test.

But in this case, since *VDD* is connected to *GND*, a relatively high current flows, which can be detected by measuring this current in a so-called *IDDQ* test [2].

2.4 Delay Faults

Delay faults are caused by small defects in the manufacturing process that cause the timing conditions of the circuits to be violated. If a delay fault occurs, the transition of a signal from 0 to 1 or from 1 to 0 does not take place in the expected time cycle but rather in the following cycle. The logical function of the circuit is not changed if the length of the clock interval is increased.

Two types of delay faults are distinguished as follows:

- *Gate delay faults*:
  A gate delay occurs if the delay is located within a single gate.
- *Path delay faults*:
  In a path delay the delay is accumulated along a path in the circuit.

When testing, delay faults can be detected by two-pattern tests applied at speed. During normal operation, delay faults can easily cause errors at the circuit outputs.

2.5 Transient Faults

Transient faults are temporary faults caused by different (often external) reasons.

Ionizing radiation of α-particles and high energy neutrons may cause very short duration pulses of current resulting in single event upsets (SEU) in the signal values of a circuit. Single event upsets may cause bits to be flipped in a latch or flip-flop, resulting in a *soft error* in the system.

Bit flips may be generated by single event upsets directly within the memory elements or in the combinational part of the circuit.

Soft errors are the main reason for errors in the memory units.

Other reasons for transient faults may be, for instance, electromagnetic fields including crosstalk between circuit lines, heat, delay faults and others. If a transient fault occurs in the combinational part of a circuit, it will be captured as a soft error in a latch or a flip-flop if the following conditions according to [16] are satisfied:

1. Logic condition:
   There must be a path which is sensitized by the input values from the location of the fault to a memory element.
2. Timing condition:
   The deviation in the logical value, which may be a short voltage pulse, has to arrive at the input of the memory element during the latching window of this element.
3. Electrical condition:
   During the propagation along the sensitized path to the memory element an electric deviation may be attenuated. The electric deviation at the location of the transient fault has to be large enough that the propagated deviation is still sufficiently large enough to be captured as an error by the memory element.

Transient faults can only be detected by concurrent checking and not be testing.

2.6 Functional Error Model

An error is caused by a fault if the input-output behavior of the faulty circuit is different from the input-output behavior of the correct circuit and if the corresponding inputs are applied to the faulty circuit.

In the following section we consider a combinational circuit f_C with the input and output sets $X = \{0,1\}^m$ and $Y = \{0.1\}^n$ where $n \geq 1$, implementing a combinational function $f : X \to Y$.

We also assume that a set $\Phi = \{\varphi_1, \ldots, \varphi_N\}$ of technical faults of the considered circuit f_C is given.

In the presence of a fault $\varphi \in \Phi$ the faulty circuit is denoted as $f_{C,\varphi}$ and the faulty circuit $f_{C,\varphi}$ implements the function f_φ, which is called an *error function*.

The notion of a functional error model is introduced by the following

Definition 2.1. The functional error model F(f) of the circuit f_C with respect to the set $\Phi = \{\varphi_1, \ldots, \varphi_N\}$ of technical faults is the set

$$F(f) = \{f_0 = f; f_{\varphi_1}, \ldots, f_{\varphi_N}\} \tag{2.4}$$

of error functions $f_{\varphi_1}, \ldots, f_{\varphi_N}$ to which the error-free function $f_0 = f$ is added.

As an example we consider the circuit $\vee_C$ consisting of a single *OR*-gate as shown in Fig. 2.1 with the set of single stuck-at faults

$$\Phi = \{\varphi_1 = 1/0, \varphi_2 = 2/0, \varphi_3 = 3/0, \varphi_4 = 1/1, \varphi_5 = 2/1, \varphi_6 = 3/1\}.$$

Then the functional error model of the *OR*-gate for the single stuck-at fault model is

$$\begin{aligned} F_{sa}(\vee) &= \{f_0 = x_1 \vee x_2; f_{1/0} = x_2, f_{2/0} = x_1, f_{3/0} = 0, f_{1/1} = f_{2/1} = f_{3/1} = 1\} = \\ &= \{x_1 \vee x_2; x_2, x_1, 0, 1\}. \end{aligned}$$

As a second example let us consider the input bridging fault of the *AND*-gate of Fig. 2.3. The set of technical faults $\Phi_{input-bridging}$ consists of the single bridging fault $brid(1,2)$ only,

$$\Phi_{input-bridging} = \{brid(1,2\}$$

If the input bridging fault $brid(1,2)$ occurs, the *AND*-gate has to be functionally replaced by an *OR*-gate and the functional error model is

$$F(\wedge)_{input-bridging} = \{x_1 \wedge x_2; x_1 \vee x_2\}.$$

If an input-output bridging fault $brid(1,3)$ as in Fig. 2.6 occurs, the erroneous circuit $\wedge_{brid(1,3)} = A_2$ is functionally equivalent to the sequential circuit represented in Fig. 2.7.

Similarly, for an input-output bridging fault $brid(2,3)$ the corresponding erroneous circuit $\wedge_{brid(2,3)} = A_3$ will also be sequential and for the *AND*-gate of Fig. 2.3 the functional error model for the technical fault model "bridging faults" Φ_{brid},

$$\Phi_{brid} = \{brid(1,2), brid(1,3), brid(2,3)\}$$

is

$$F(x_1 \wedge x_2)_{brid} = \{x_1 \wedge x_2; x_1 \vee x_2, A_2, A_3\}.$$

In this case the functional error model consists of a set of error functions, in this case a single function $x_1 \vee x_2$, and a set of error automata, A_2 and A_3.

The error model for a *CMOS* implementation of a combinational circuit with stuck-on and stuck-open transistor faults consists of a set of error functions and a set of error automata.

A different situation occurs if, as in Fig. 2.8, an inverter is included in the feedback path of a bridging fault. Then, as already described, the circuit output of the faulty circuit will be oscillating.

Such oscillating behavior cannot be described by use of the classical discrete system theory as a combinational function or as an abstract automaton. In this case a functional error model cannot be formulated as described.

Another limitation of this model results from the fact that not all possible faults can be formally specified.

As long as the faulty circuit can be adequately described as a combinational function or an abstract automaton the functional error model is a precise description of the errors expected due to the technical faults considered.

2.7 Output Dependencies

Output dependencies with respect to technical faults can be successfully used for the design of error detection circuits. This approach will be shown in this book.

Until now *independent* outputs, *weakly independent* outputs and *unidirectionally independent* outputs were investigated. Structural and functional dependencies of outputs may be distinguished.

Independent outputs will be discussed first of all.

Definition 2.2. Two outputs y_1 and y_2 of a circuit f_C with the input set X are regarded as independent with respect to a fault φ if, in the presence of the fault φ for every input $x \in X$ at most one of the outputs y_1 or y_2 is erroneous [17].

Definition 2.3. A group of outputs $y_1, y_2, \ldots, y_n$ of a circuit f_C with the input set X is regarded as independent with respect to a fault φ if, in the presence of the fault φ for every $x \in X$ at most one of these outputs is erroneous.

If the outputs $y_1, y_2, \ldots, y_n$ are implemented without sharing common gates, they are (structurally) independent with respect to single stuck-at faults.

Obviously a group of independent outputs can be successfully monitored by parity prediction and several groups of independent outputs by group parity prediction.

Since groups of independent outputs seldom as a generalization comprise independent outputs, the notion of weakly independent outputs was introduced. As a modification of Definition 2.2 we have

Definition 2.4. Two outputs y_1 and y_2 of a circuit f_C with the input set X are regarded as weakly independent with respect to a fault φ if, in the presence of the fault φ there exists an input $x_\varphi \in X$ that for this input x_φ one of the outputs y_1 or y_2 is erroneous [18].

Groups of weakly independent outputs are similarly defined. If groups of weakly independent outputs are monitored by parity prediction faults may be detected by some latency.

Groups of independent and weakly independent outputs can be successfully utilized for the design of self-checking and self-testing circuits. This will be described in Section 3.3.4 of this book.

Also unidirectionally independent outputs are considered.

Definition 2.5. Two outputs y_1 and y_2 of a circuit f_C with the input set X are called unidirectionally independent with respect to a fault φ if, in the presence of the fault φ for every input $x \in X$ either both the outputs are correct, or only one of the outputs is erroneous, or, if both the outputs are erroneous they are unidirectionally erroneous.

In the last case both the outputs are erroneously changed from $0,0$ to $1,1$ or from $1,1$ to $0,0$ but not from $1,0$ to $0,1$ or from $0,1$ to $1,0$.

Groups of unidirectionally independent outputs which can be adequately checked by Berger codes were introduced in [19].

In [19] a simple transformation is also described that transforms an arbitrarily given combinational circuit into a circuit with unidirectionally independent outputs. This circuit transformation is very similar to the transformation which is used to transform a self-dual circuit into a self-dual fault-secure circuit as described in Section 3.8.4 of this book.

2.8 Self-Testing and Self-Checking

Errors of combinational or sequential circuits which are caused by technical faults which are not detected by testing shall be detected during normal operation by concurrent checking.

The necessary error detection probability is determined by the type of application and by the probability of the occurrence of an error of the monitored circuit during normal operation.

Let us assume that a functional circuit produces (on average) an error every 10 days. If the error detection probability is 90%, we can expect a single undetected error within 100 days.

If the monitored functional circuit produces a single error per day (per 144 minutes = $1/10$ day), then an error detection probability of 99% (99.9%) is needed to achieve the same expected number of one undetected error in 100 days.

Clearly the required error detection probability depends very much on the type of application, which is different for a cell phone, a PC, an air plane or a nuclear power station.

However, it is clear that 100% error detection probability can never be achieved. And also due to the nature of probability even a very high error detection probability never excludes a single event of an undetected error among a huge number of error free outputs and a large number of detected errors. It is not predictable in what moment an undetected or undetectable error occurs.

To classify different levels of error detection the notions of *self-testing*, *fault-secure*, *totally self-checking* and *code-disjoint* are mainly used. These notions will be given now for a circuit C with concurrent checking, with an input set X, a subset $\chi = X_{exp}$ of expected inputs, $\chi \subset X$, for which the behavior of C is of interest and a set of faults $\Phi = \{\varphi_1, \ldots, \varphi_N\}$:

- *Self-testing*:
  A circuit C with concurrent error detection is self-testing with respect to a set Φ of faults if for every fault $\varphi \in \Phi$ there exists an input $x_\varphi \in \chi$ such that for the input x_φ the output of C is erroneous and the error will be detected by concurrent checking.
- *Fault-secure*:
  A circuit C with concurrent error detection is fault-secure with respect to a set Φ of faults if every error at the output of C due to every fault $\varphi \in \Phi$ for every input $x, x \in \chi$ will be detected by concurrent checking.
  In a fault-secure circuit every erroneous output which is caused by a fault of the considered fault model will always be detected by concurrent checking.
- *Totally self-checking*:
  A circuit C with concurrent error detection is totally self-checking with respect to a set Φ of faults if C is self-testing and fault-secure with respect to Φ.

Very often the notions self-testing, fault-secure and totally self-checking are defined with respect to the fault set of all single stuck-at faults of C where the circuit C also includes the hardware for concurrent checking.

It is clear also from the definitions of self-testing, fault-secure and totally self-checking with respect to a given set Φ of faults that there may be errors caused by faults, not from the set Φ, which may not be detected. This is, for instance, possible if the circuit is totally self-checking with respect to all single stuck-at faults but not for all bridging faults.

The detection of input errors is a particular problem of concurrent checking. Without special measures an erroneous input of a circuit with concurrent checking is interpreted by the circuit simply as a different input for which no error indication is demanded.

Input errors have to be detected if they are not detected as output errors of the preceding system.

To detect input errors, the set X of inputs has to be divided into a subset set of expected inputs X_{exp} and the remaining set of unexpected inputs X_{nexp} with $X = X_{exp} \cup X_{nexp}$, and $X_{exp} \cap X_{nexp} = \emptyset$. The expected inputs have to be from the expected input set $X_{exp} = \chi$.

As a subset of X the set X_{exp} is called a code, and the expected inputs - which are elements of X_{exp} - are called input code words.

As an example, let us consider a circuit C with m dimensional inputs $x = x_1, x_2, \ldots, x_m$. The set X_{exp} of expected inputs may be the set of all m dimensional binary vectors with even parity. Then the expected inputs are code words of an even parity code.

In general any other subset of m dimensional inputs may be a set of expected inputs.

It is checked whether the inputs belong to the set of expected inputs X_{exp} or not, and an input error can be detected if it changes an expected input $x \in X_{exp}$ into an erroneous input $x_{err} \in X_{nexp}$. If an error changes an expected input $x \in X$ into another input x', $x' \neq x$, $x' \in X_{exp}$ then the error cannot be detected.

If, as previously assumed, the set of expected input values is the set of all m dimensional binary vectors with even parity, then an input error will be detected if the parity of the erroneous input is odd and the input error will not be detected if the parity of the erroneous input is even.

To also include the detection of input errors in the concept of concurrent checking, the notion of *code-disjoint circuits* was introduced:

- *Code-disjoint*:
  A circuit C with concurrent checking is code-disjoint with respect to a code $\chi = X_{exp} \subset X$ if, for an erroneous input $x_{err} \neq x$, where $x \in X_{nexp}$, an error is detected by concurrent checking and if, for $x_{err} \in X_{exp}$, no error is indicated.

For the design of self-checking circuits the set of single stuck-at faults is used as the set of technical faults Φ in almost all cases. There are different reasons for the choice of this set of faults, as given below:

1. Almost all known methods for the design of self-checking circuits are based on the fault model of single stuck-at faults.
2. There is a strong belief in the community that a circuit which is fault-secure for all single stuck-at faults and detects all errors due to single stuck-at faults will also detect most of the errors caused by other, even non-modelled, faults.
   If the circuit is fault-secure, then for an arbitrary stuck-at fault, a fault will be detected for every input which generates an erroneous output.
3. The method of n-detection testing has some similarity to the concept of totally self-checking. n-detection of single stuck-at faults is also believed to be a method to detect non-modelled faults by testing. In the terminology of testing totally self-checking means that every single stuck-at fault will be detected for every possible test of this fault by the hardware of concurrent checking. Thus, if there exist n test vectors for a single stuck-at fault φ, all these n test vectors of an n-detection test are utilized to detect all single stuck-at faults by concurrent checking.

For sequences of (possibly accumulating) faults the notions of strongly fault-secure and strongly code-disjoint are given in [5, 20, 21].

2.9 Faults and Errors in Submicron Technologies

With the fast development and application of the submicron technologies concurrent checking is becoming increasingly important. This is due to the fact that in these technologies chips cannot be tested with the same high fault coverage of 99.x% for their permanent faults as for single stuck-at faults and that the number of transient faults is increasing [22, 23].

The main reasons, which will be briefly described, are the large chip sizes, the very small dimensions of the transistors, the shrinking voltage levels and the lack of simple adequate fault models for the new types of faults.

1. Large chip sizes:
   It is not feasible to test very large chips of several tens of millions of transistors for all possible modeled and unmodeled hundreds of millions of permanent faults within a few seconds in a high volume production test.
2. Small dimensions of the transistors:
   The dimensions of the transistors and of the connecting wires are so small that statistical variations may already result in subtle defects, such as high-resistance bridging faults or high-resistance opens causing very small delays. These small delays may show up as erroneously delayed output signals only at the high clock frequencies used in normal operation, and not in the test mode.
3. Growing number of transient faults:
   The number of transient, i.e. of non-permanent faults caused by α-particles and by cosmic radiation is growing due to the small dimensions of the transistors and due to the reduced voltage levels. Also crosstalk errors and delay faults due to subtle defects cannot be modeled other than by random transient faults.
4. Non-existing simple fault models:
   For the design of effective tests, simple formalized fault models, and test metrics based on such models are needed. Such simple fault models and test metrics do not exist for most of the realistic faults. Therefore, in practice, for the design of effective tests, besides the modeled single stuck-at faults, bridging faults, stuck-on faults, stuck-open faults and other modeled faults, also *unmodeled faults* have to be taken into account. Since these unmodeled faults are not well-described no guarantee can be given that they are detected by the test.

At the end of this chapter we refer to some recent attempts, to solve the problems of concurrent checking in submicron technologies by use of different additional sensors on the chips. As examples we mention here:

- In [24] inverter chains are applied to measure the amount of radiation to which the circuit is exposed. In the case of a high level of radiation, the output capacities of the circuit are temporarily increased to make the circuit less sensitive to single event upsets.
- In [25] the detection of aging faults is considered. These faults occur as delay faults after the chips are used for some years. These faults cannot be detected by a production test. During the test they do not yet exist.
  Special aging sensors are proposed to *predict* possible delay faults before they actually appear. The clock rate of the chip will be adaptively slowed down if necessary.
- In [26] embedded sensors are applied to detect the silicon ionization and to predict single event upsets in an early stage of their generation.

The development and application of different sensors for concurrent checking in the submicron technologies is in a very early stage, and the part such additional sensors will play in the future is not easy to predict. The application of the proposed sensors reduces the probability of special types of undetected errors in the submicron technologies. With the reduced rates of undetected errors the well developed logical methods of concurrent checking, as described in this book at the gate level, will be become more effective.

We expect that within the next 5–10 years that the combination of the logical methods of concurrent checking with some future sensors will be an active area of research.

In this chapter the most relevant types of technical faults, namely single stuck-at faults, bridging faults, stuck-open and stuck-on transistor faults, delay faults, transient faults in the combinational part of a circuit and directly induced soft errors in the memory elements were described.

It was shown how these types of faults influence the behavior of the faulty circuit and how the corresponding errors can be adequately modeled.

The functional error model corresponding to a set of physical faults was presented. It was explained that the functional error model is a very general and exact error model as long as the faulty circuit can be described as a combinational or sequential time-discrete circuit and has neither oscillating signal lines nor undefined values.

Independent outputs, weakly independent outputs and unidirectionally independent outputs were introduced to model the different dependencies between pairs or groups of circuit outputs in the presence of faults. These dependencies will be utilized in the following chapters for the design of error detection circuits.

To qualify the error detection capability of a circuit with concurrent checking, the definitions of self-testing, fault-secure, totally self-checking and code-disjoint circuits with respect to a given fault set were presented. These definition were formulated for circuits with concurrent checking.

The fact that these notions are mainly defined in the literature with respect to the fault model of all single stuck-at faults was intensively discussed.

Chapter 3
Principles of Concurrent Checking

3.1 Duplication and Comparison

In this section duplication and comparison as the conceptually simplest method of error detection will be described and discussed in detail.

The outputs of two duplicated circuits are compared during normal operation by a comparator or, if the circuits are inversely duplicated, by a two-rail checker.

The main advantages of this method are that the method is simple, that it is easily applicable to every circuit and that no specific error model is needed.

The main disadvantages are the high area overhead and the more than two-fold increase in power consumption.

Of special interest is how the potential faults within the comparator can be detected. It will be shown how this problem can be traditionally solved by use of self-checking comparators. Self-checking comparators with two outputs and with a single periodic output will be presented. These comparators are able to detect all internal single stuck-at faults.

A new easily testable comparator with a single output will also be described. It will be shown how the most disturbing faults, stuck-at faults at the single output of the comparator, can be tested without interrupting the normal operation of the comparator.

Partial duplication combining error detection by duplication and comparison for the duplicated part and by parity prediction for the non-duplicated part will be explained. The method of partial duplication will be utilized for the design of self-checking adders in Chapter 4 of this book.

The necessary area can be reduced by the method of partial duplication and all errors due to faults within the duplicated part of the circuit, including odd and even errors within the output registers, are detected by comparing the contents of these registers. It is of special interest that these registers are checked by duplication and comparison, since soft errors directly induced in these registers by α-particles are of arbitrary parity and cannot be detected by parity prediction.

The errors caused by faults within the non-duplicated part are in most cases single-bit errors detected by parity prediction.

Duplication and comparison are the standard method for error detection, and all the other methods will be compared with this method with respect to the necessary area, power consumption and error detection probability of errors due to single stuck-at faults.

3.1.1 Description of the Method

The method of duplication and comparison is shown in Fig. 3.1. The monitored circuit C^1 is functionally duplicated in the circuit C^2. The circuits C^1 and C^2 are functionally, but not necessarily structurally, equivalent.

During normal operation an input sequence $x_1, x_2, \ldots, x_i, \ldots$ of m dimensional inputs is applied to both the circuits C^1 and C^2. The corresponding n dimensional outputs $y_1^1, y_2^1, \ldots, y_i^1 \ldots$ of C^1 and $y_1^2, y_2^2, \ldots, y_i^2 \ldots$ of C^2 are compared by a comparator *Comp*. The error signals $e_1, e_2, \ldots, e_i, \ldots$ are the output signals of the comparator *Comp*.

For $y_j^1 = y_j^2$ the error signal e_j is $e_j = 0$ and no error is indicated. For $y_j^1 \neq y_j^2$ the error signal e_j is $e_j = 1$, which indicates an error.

If, instead of the directly duplicated circuit C^2, an inverted duplicated circuit $\overline{C}^2$ is then designed for $j = 1, 2, \ldots, n$ the inverted outputs $y_j^2 = \overline{y}_j^1$ of C^1 are implemented by $\overline{C}^2$, and the outputs of C^1 and $\overline{C}^2$ have to be compared by a two-rail checker *TRC*. This is shown in Fig. 3.2.

Advantages of duplication and comparison are as follows:

1. The method is simple.
2. Every error at the outputs of one of the duplicated circuits C^1 or C^2 is immediately detected.
3. No special fault model is required.

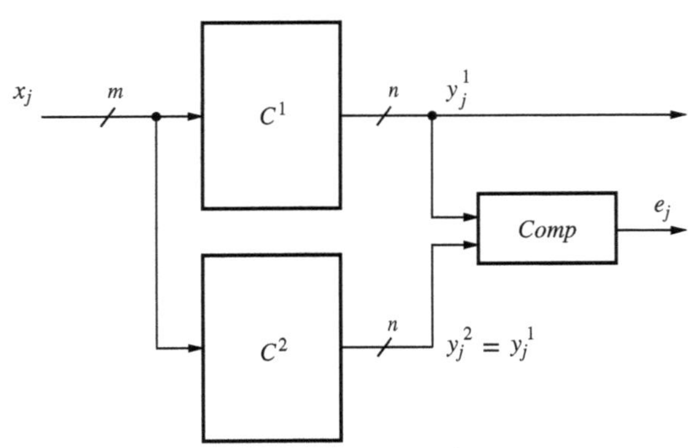

Fig. 3.1 Duplication and comparison

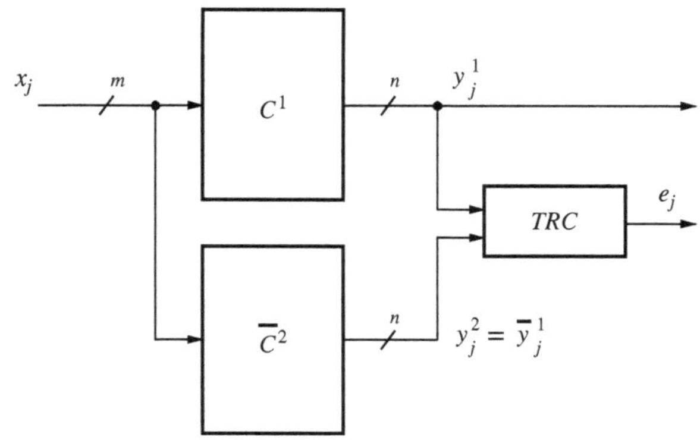

Fig. 3.2 Inverted duplication

Disadvantages of duplication and comparison are as follows:

1. The area required is more than twice the area of the original circuit C^1.
   The original circuit C^1, the duplicated circuit C^2 and the comparator *Comp* have to be implemented. An area of about 220–250% of the area of the original circuit is needed for the implementation of duplication and comparison.
2. The power consumption is also more than twice the power consumption of the original circuit C^1.
3. The number of expected faults of the circuitry of Fig. 3.1 or Fig. 3.2 is more than twice the number of expected faults of the original circuit C^1.
   A larger number of faults causes larger yield losses in production.
4. Faults within the comparator *Comp* or the two-rail checker *TRC* can prevent other faults from being detected.

3.1.2 Comparators and Two-Rail Checkers

The possibilities for the detection of faults in the comparator or in the two-rail checker are now discussed in more detail.

If both the circuits C^1 and C^2 are correct, the outputs y_i^1 of C^1 and y_i^2 of C^2 are equal for $i = 1, 2, \ldots$. Since the word length of $y_i^1 = y_i^2$ is n at most 2^n of the principally possible 2^{2n} input values can be applied to the $2n$ inputs of the comparator *Comp*.

Thus for $n = 10$ of the $2^{2n} = 2^{20} \approx 10^6$ possible input values of the comparator as a circuit with 20 input lines only $2^n = 2^{10} \approx 10^3$ different values may be applied.

In reality even a small subset of these 2^n principally possible values may be generated at the outputs of C^1 and C^2. The same is true for two-rail checkers. Because of this relatively very small set of input values which are applied during normal operation to the inputs of the comparator *Comp* the detection of faults in the comparator during normal operation is a serious problem. The same is true if an embedded comparator has to be tested. Undetected faults may be accumulated within the comparator preventing errors at its inputs from being detected.

Solutions to this problem were investigated for a long time.

We now discuss the problem for two-rail checkers in greater detail. (By placing n inverters to n of the $2n$ inputs of a two-rail checker the two-rail checker can be easily modified in an equality checker.)

A well known two-rail checker with four inputs according to [27] is shown in Fig. 3.3. Two-rail checkers with more than four inputs can be obtained as trees of such two-rail checkers with four inputs and two outputs.

The two-rail checker of Fig. 3.3 maps all two-rail inputs (i.e. the outputs of C^1 and $\overline{C}^2$) with $y_1^1 = \overline{y}_1^2$ and $y_2^1 = \overline{y}_2^2$ to two-rail outputs r_1, r_2 with $r_1 = \overline{r}_2$. All inputs that are not two-rail are mapped to equal outputs $r_1 = r_2$. The two-rail checker of Fig. 3.3 has the four correct two-rail input words 0101; 0110; 1001; 1010 which are called the input code words of the checker. The correct two-rail output words of this checker are 01; 10. They are called the output code words of the checker.

All the input code words 0101; 0110; 1001; 1010 are mapped to the two output code words 10; 01 and all the input non-code words 0000; 0001; 0010; 0011; 0100; 0111; 1000; 1011; 1100; 1101; 1110; 1111 are mapped to one of the output non-code words 00 or 11.

The property of a checker to map input code words to output code words and input non-code words to output non-code words is called *code-disjointness* [28]. The property of a checker to be code-disjoint is necessary to detect all errors at its inputs.

The detection of internal faults of a checker is, as already pointed out, also of great interest. If internal faults of a checker are not detected in time these undetected faults may destroy the code-disjointness of the checker.

Faults of the checker can be either detected by concurrent checking during normal operation or by testing in a special test mode.

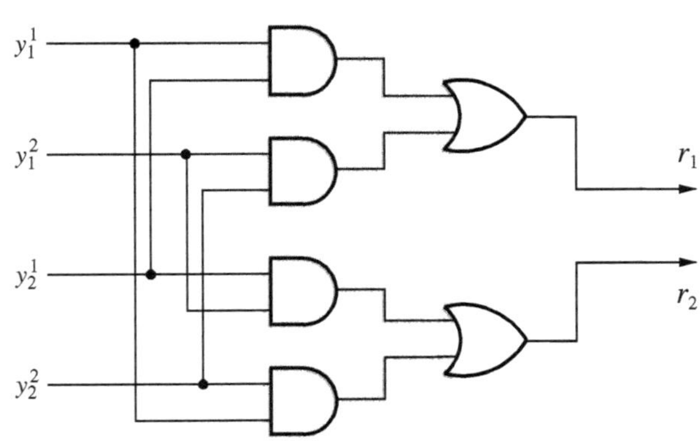

Fig. 3.3 Self-checking two-rail checker

A checker is called *self-testing* if for every (single stuck-at) fault φ there exists an input code word x_φ such that in the presence of the fault φ the faulty checker outputs a non-code output for the input x_φ.

The internal fault φ is detected if the input code word x_φ is actually applied to the checker during normal operation. If the input code word x_φ is in the set of expected inputs of the checker then we can expect that the fault x_φ will be detected within a reasonable time interval. If x_φ does not belong to the set of expected inputs of the (embedded) checker, the input code word x_φ never occurs as an input of the checker and the fault φ will not be detected.

If for the two-rail checker of Fig. 3.3 all the four possible two-rail input code words 0101; 0110; 1001; 1010 are actually applied the, two-rail checker is self-testing with respect to all single stuck-at faults. This can be easily proven by inspection.

Thus the checker of Fig. 3.3 is code-disjoint and, if all the possible four input code words are actually applied to its inputs, also self-testing (with respect to single stuck-at faults).

If only a single output of the checker is allowed, the outputs of the two-rail checker in Fig. 3.3 have to be XORed. But in this case a single stuck-at fault at the single output line of the checker cannot be detected during normal operation.

Some of the single stuck-at faults within the checker may remain undetected if, during normal operation, not all the possible two-rail inputs are applied to the inputs of a two-rail checker. This may be often the case if real circuits with a large number of outputs are (inverted) duplicated. In the presence of undetected stuck-at faults within the checker some of the input non-code words are now mapped to output code words, and they cannot be detected as erroneous. The faulty checker loses the property of being code-disjoint and some of the errors at its inputs cannot be detected.

There are many attempts to design self-checking two-rail and equality checkers or comparators. We mention here only some results concerning comparators with a single dynamic periodic output [29, 30, 31].

In Fig. 3.4 a self-checking comparator with a single periodic output according to [29] is shown. The comparator of Fig. 3.4 consists of an array of four XOR-gates and a *C-element* with two inputs according to [32]. The comparator componentwise compares $y^1 = (y_1^1, y_2^1))$ and $y^2 = (y_1^2, y_2^2)$. The additional input signal y_0 is periodically changed between 0 and 1 and the output v is periodic as long as no error occurs.

For $y_0 = 0$ we have

$$z_1 = z_2 = 0 \quad \text{for} \quad y_1^1 = y_1^2 \quad \text{and} \quad y_2^1 = y_2^2$$

and for $y_0 = 1$

$$z_1 = z_2 = 1 \quad \text{for} \quad y_1^1 = y_1^2 \quad \text{and} \quad y_2^1 = y_2^2.$$

If the additional signal y_0 is periodic and if no error occurs, the internal signals z_1, z_2 with $z_1 = z_2$ are changing periodically between 0,0 and 1,1. These

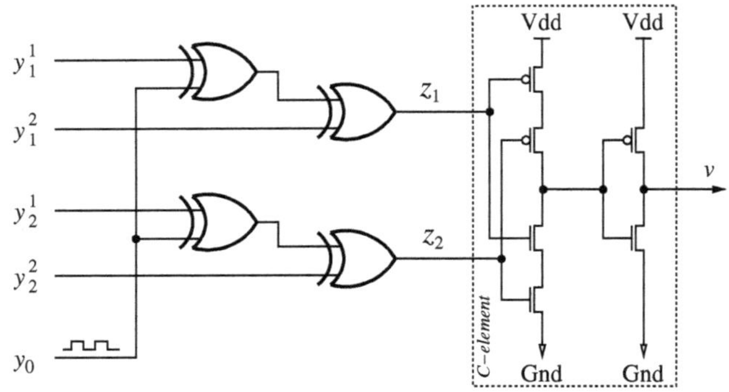

Fig. 3.4 Self-checking two-rail checker with a single periodic output

periodic internal signals are processed by the *C-element* into a single periodic output $1,0,1,0,1,\ldots$.

The function of the *C-element* is described in Table 3.1.

If, due to an error at one of the inputs, say at the input y_1^2, we have for the first time $z_1 \neq z_2$, the corresponding output v of the *C-element* is the previous output of this element. Then the output is not periodic and the error will be detected. Also in the case that due to an error both the components of the inputs are not equal the output v of the *C-element* is not periodic.

The comparator of Fig. 3.4 can be tested by just two inputs. The described comparator can be also used as a two-rail checker.

A comparator is described in [29] for more than two inputs. The output of the described comparator is, as already pointed out, a dynamic output and a *C-element* is needed which may not belong to a commercial synthesis tool.

Instead of designing a self-checking comparator it is also reasonable to design a comparator which is periodically tested.

For practical applications we recommend the easily testable comparator with a single output according to [33], which will now be described. Only a standard library is needed and the comparator can be implemented by any design tool. In contrast to the known comparators with a single output [29, 30, 31], the output of the comparator of Fig. 3.5 is static and not dynamic.

Table 3.1 Function of *C-element*

z_1	z_2	v
0	0	0
1	1	1
0	1	previous value
1	0	previous value

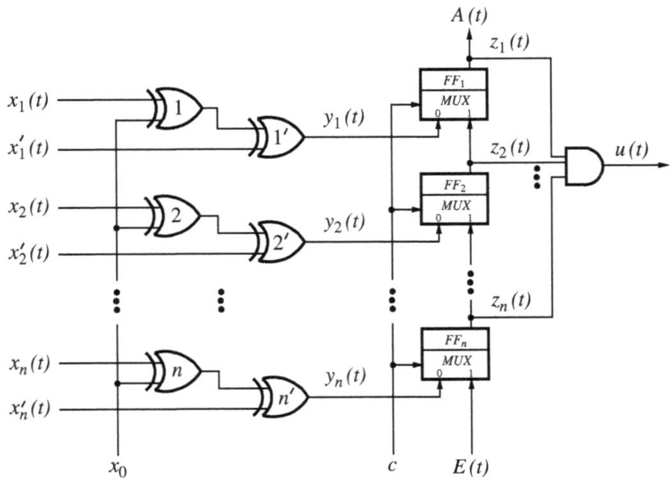

Fig. 3.5 First self-testing comparator with a single output

The comparator shown in Fig. 3.5 compares the two input vectors $x_1(t),\ldots,x_n(t)$ and $x'_1(t),\ldots,x'_n(t)$. For the additional input $x_0 = 1$ the comparator is an equality checker and for $x_0 = 0$ a two-rail checker.

The comparator is not totally self-checking, but compared to the comparator used in [34], can be easily tested.

The comparator of Fig. 3.5 consists of three serially connected parts. The first part is a simple XOR-network with $2n$ XOR-gates. The second part consists of n scannable flip-flops, which can be used either as a pipeline register or as a scan path. The third part is an n-input *AND (NAND)*-gate.

For $i = 1,\ldots,n$ the input signal $x_i(t)$ and the additional input x_0 are connected to the inputs of the XOR-gate XOR_i. The output of this XOR-gate and the input signal $x'_i(t)$ are connected to the inputs of the XOR-gate XOR'_i, the output of which is denoted by $y_(t)$.

$y_i(t)$ is determined as

$$y_i(t) = x_i(t) \oplus x'_i(t) \oplus x_0. \tag{3.1}$$

If the control signal c of the multiplexors of the scan flip-flops is 0, then $y_i(t)$ is stored for one clock cycle in the flip-flop FF_i and the output $z_i(t)$ of FF_i is

$$z_i(t) = y_i(t-1) = x_i(t-1) \oplus x'_i(t-1) \oplus x_0. \tag{3.2}$$

Then the output $u(t)$ of the comparator is

$$u(t) = \bigwedge_{i=1}^{n} y_i(t) = \bigwedge_{i=1}^{n} (x_i(t-1) \oplus x'_i(t-1) \oplus x_0). \tag{3.3}$$

For $x_0 = 1$ the comparator is an equality checker. As long as all components of the input vectors $x_1(t),\ldots,x_n(t)$ and $x'_1(t),\ldots,x'_n(t)$ are pairwise equal the output $u(t+1)$ is equal to 1. If at least one of the components $x_k(t)$ and $x'_k(t)$ are not equal, $u(t+1)$ is equal to 0 and every erroneous pair of inputs will be detected at the output of the comparator.

For $x_0 = 0$ the comparator is a two-rail checker. As long as all components of the input vectors are pairwise different the output of the comparator is equal to 1. If at least one pair $x_k(t), x'_k(t)$ of the components of the input vector is equal, the output $u(t+1)$ is equal to 0.

Now we show how the comparator with $x_0 = 1$ can be tested with respect to all single stuck-at faults. With $x_0 = 1$ the comparator is an equality checker.

First we consider the testability of the XOR-gates. A two-input XOR-gate can be tested with respect to single stuck-at faults if all the possible inputs 00, 01, 10 and 11 are applied to its inputs. During normal operation we have $x_i(t) = x'_i(t) \in \{00, 11\}$ and $x_0 = 1$ and the inputs of the XOR-gates are 01 and 11 for XOR_i and 10 and 01 for $XOR`_i$. If we change the additional input $x_0$ to 0, then the inputs of the XOR-gates are 00 and 10 for $XOR_i$ and 00 and 11 for $XOR`_i$. For $x_0 = 0$ and for $i = 1,\ldots,n$ the internal signals $y_i(t)$ are stored in the flip-flops FF_i and shifted out of the scan chain.

During normal operation the n-input AND-gate at the output of the comparator is tested with respect to all stuck-at-0 faults. The AND-gate can be easily tested for stuck-at-1 faults by use of the scan chain. The necessary n test-inputs are shifted in and the output is monitored.

A modification of the comparator of Fig. 3.5 is shown in Fig. 3.6.

The comparator shown in Fig. 3.6 allows us to check the single output line for a single stuck-at fault during normal operation.

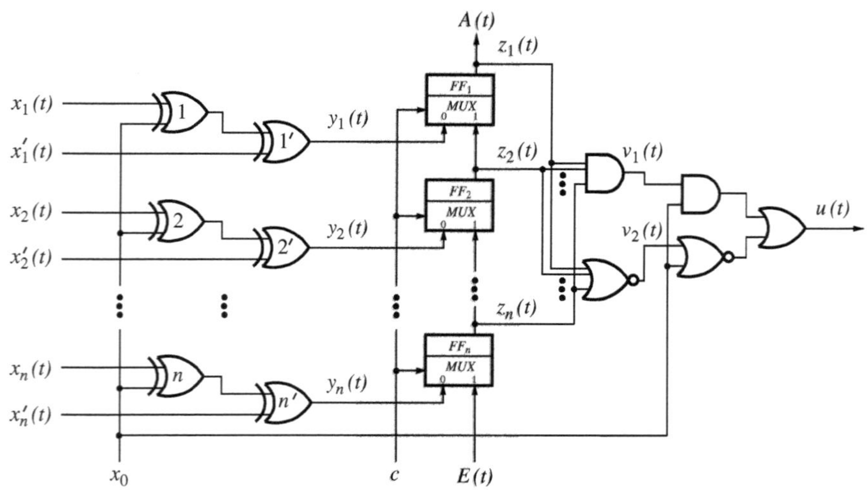

Fig. 3.6 Second self-testing comparator with a single output

In Fig. 3.6 the outputs of the flip-flops $FF_1,\ldots,FF_n$ are connected to both an n-input AND-gate with the output

$$v_1(t) = z_1(t) \wedge \ldots \wedge z_n(t)$$

and an n-input NOR-gate with the output

$$v_2(t) = \overline{z_1(t) \vee \ldots \vee z_n(t)}.$$

The additional input x_0 of the XOR-gates $XOR_1,\ldots,XOR_n$ also controls whether $v_1(t)$ or $v_2(t)$ is selected as the output $u(t)$ of the comparator. Thus we have

$$u(t) = x_0 v_1(t) \vee \overline{x_0 \vee v_2(t)}. \tag{3.4}$$

For $x_0 = 1$ the output of the comparator is

$$u(t) = \bigwedge_{i=1}^{n} (x_i(t-1) \oplus x_i'(t-1) \oplus 1), \tag{3.5}$$

and for $x_0 = 0$

$$u(t) = \bigvee_{i=1}^{n} (x_i(t-1) \oplus x_i'(t-1)). \tag{3.6}$$

If no error occurs, equations (3.5) and (3.6) imply for an equality checker

$$u(t) = \begin{cases} 1 \text{ for } x_0 = 1 \\ 0 \text{ for } x_0 = 0 \end{cases}.$$

By changing the additional input x_0 from 1 to 0 the output of the comparator can be tested for stuck-at 0/1 faults. Consequently, the function of the comparator to check its input signals for equality is not interrupted.

All single stuck-at faults of all gates, except the single stuck-at-1 faults at the inputs of the n-input AND-gate and the single stuck-at-0 faults at the inputs of the n-input NOR-gate, will be detected during normal operation if the additional input x_0 is switched from time to time between 0 and 1.

The single stuck-at-1 faults at the inputs of the n-input AND-gate and the single stuck-at-0 faults at the inputs of the n-input NOR-gate can be tested in a test mode by scanning the necessary test inputs into the scan chain.

If the comparator is slightly modified, it can be beneficially applied for error location in fault-tolerant systems. Details are described in [33].

A well-known modification of duplication and comparison is *Two-Rail Logic*.

In Two-Rail Logic every circuit line of the original circuit is duplicated in two lines. The first of these duplicated lines carries the original signal s and the second one the corresponding inverted signal $\overline{s}$. Inverters can be saved since both the original and the inverted signal are always available. A disadvantage of this method is that the circuit cannot be simply duplicated, a special design for the two-rail circuit is necessary. Two-rail logic is described, for instance, in [35, 36].

3.1.3 Method of Partial Duplication

Error detection by *partial duplication* was introduced in [37] and in recent years applied to different types of adders, multipliers and dividers in [38, 39, 40, 8, 9, 10].

We consider a combinational circuit C, the outputs of which are stored in a latch or a register R. It is assumed that the considered circuit C is implemented as a serial connection of two circuits C_1 and C_2, as shown in Fig. 3.7.

Such an implementation can be easily obtained from the netlist of C by splitting the netlist into two parts.

The sub-circuit C_2 and the register R are duplicated in the two sub-circuits C_2^1 and C_2^2 and the two registers R^1 and R^2. If an error occurs due to a fault in one of the duplicated circuits C_2^1 and C_2^2 or in one of the registers R^1 or R^2, the error will be detected by comparing the contents of the registers R^1 and R^2. The combinational circuit C_1, which is not duplicated, is monitored by an error detection circuit. Figure 3.8 shows a partially duplicated circuit where the non-duplicated part C_1 is checked by parity prediction. The outputs of C_1 are XORed to derive the output parity $P(z)$ of C_1 and compared with the predicted parity $P_z(x)$, which is determined from the inputs x of C_1.

Soft errors directly induced in the registers R^1 or R^2 and soft errors in the registers caused by transient faults in the duplicated parts C_2^1 or C_2^2 near to the registers are detected by comparing the contents of the registers. Errors due to faults in the non-duplicated part C_1 result in the same erroneous contents of both the registers R^1 and R^2. They cannot be detected by comparing the contents of these duplicated registers. These errors are detected by the error detection circuit for C_1.

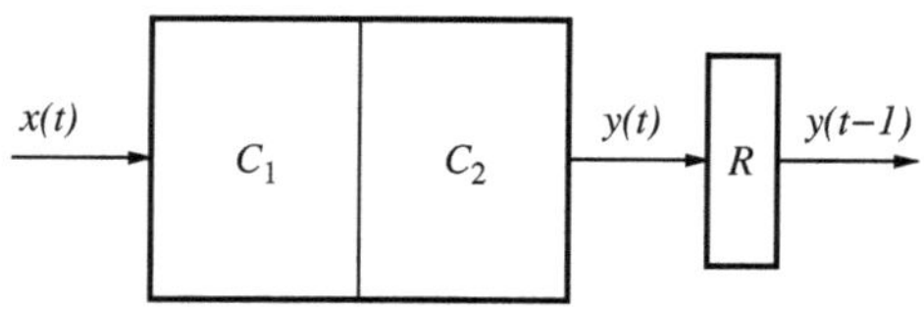

Fig. 3.7 Circuit represented as a serial connection of C_1 and C_2

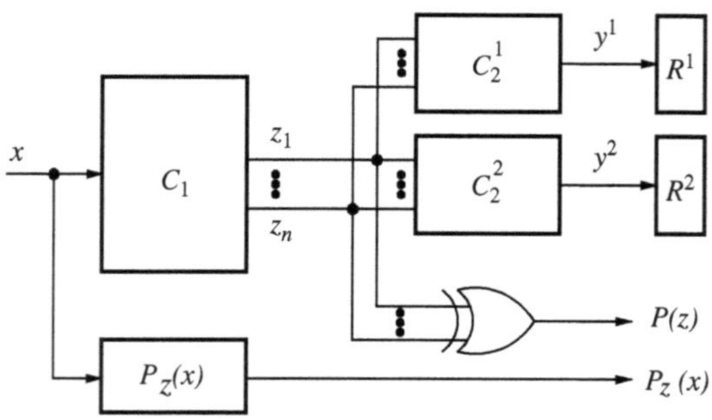

Fig. 3.8 Partial duplication with parity checking for the non-duplicated part

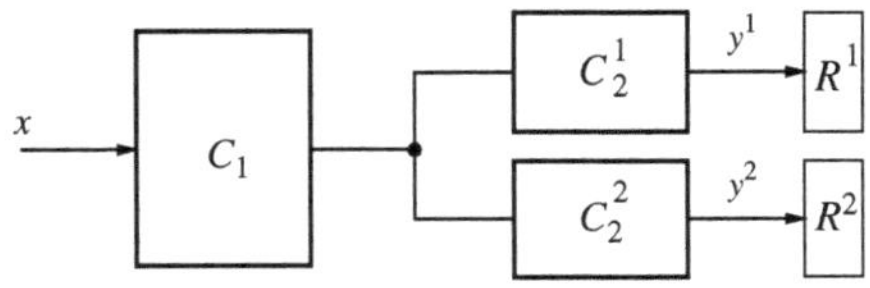

Fig. 3.9 Partial duplication without concurrent checking for the non-duplicated part

Code-disjoint partially duplicated circuits were introduced in [41] and described in Section 3.5 of this book.

Partial duplication without error detection for the non-duplicated combinational part C_1 according to [42] is shown in Fig. 3.9.

The duplicated circuit C_2 is determined such that the gates with a high soft error susceptibility are added step by step to C_2 until the necessary area for the implementation of C_2 does not exceed the acceptable additional area. Faults within C_2^1 and C_2^2 will be detected, while faults within C_1 will not be detected. Since the gates with high susceptibility to soft errors belong to C_2^1 and C_2^2, the probability of error detection with respect to soft errors is high.

This section explained in detail duplication and comparison as the conceptually simplest method of error detection.

The main advantages of this method - namely that the method is easily applicable, that no specific error model is needed and that every error due to a fault in one of the duplicated circuits will be detected were described.

Also the main disadvantages, the necessary large area, which is more than twice the area of the functional circuit, and the high power consumption, which is also more than twice the power consumption of the original circuit, were discussed.

It was explained that faults in the comparator that can prevent the comparator from detecting errors of the functional circuits require special treatment and that these errors can be detected by self-checking comparators. Self-checking comparators with two outputs and a single periodic output were presented.

A new easily testable comparator with a single output and an additional input signal was also introduced. It was shown how the most disturbing faults, stuck-at faults at the single output of that comparator, can be tested without interrupting the normal operation of the comparator.

Partial duplication with error detection by duplication and comparison for the duplicated part and error detection by parity prediction for the non-duplicated part were described. It is of special interest that the output registers were included in the duplicated part since soft errors directly induced by α-particles in these registers are of arbitrary parity and cannot be detected by parity prediction.

Errors caused by faults in the non-duplicated part are in the most cases one-bit errors and are detected by parity prediction.

Partial duplication without error detection for the non-duplicated part was likewise considered.

Since duplication and comparison are the standard method for error detection, all the other methods described in this book will be compared with this method with respect to the necessary area, power consumption and error detection probability of errors due to single stuck-at faults.

3.2 Block Codes for Error Detection

In this section the basic notion and notations of block codes will be described. It will be explained how the check bits are determined from the information bits, and the classical codes which are utilized for the design of error detection circuits will be briefly described. Almost all of these codes are systematic block codes.

A new class of codes, non-linear split error detection codes, where parities are split into non-linear parts will be introduced. The error detection capability of these codes will be investigated and it will be demonstrated that a first subset of errors will be detected with certainty, a second subset of errors with a probability greater or equal to $1/2$ and a third very small subset of errors will not be detected. This is different to linear codes, where an error is either detected with certainty or not detected with certainty.

The expenditure for the implementation of these codes lies somewhere between the expenditure required for parity codes and the expenditure required for Hamming codes.

This section will not be an introduction to coding theory and we assume that the reader is familiar with the basics of coding theory as described in the literature for coding theory, for instance, in [43, 44].

3.2.1 Classical Error Detection Codes

In general, a code is a subset ψ of a set X, $\psi \subseteq X$. The element x, $x \in X$, is an element of the code or a code word if $x \in \psi$ and a non-code word if $x \in X \backslash \psi$.

A well-known example of a code is the parity code. For a parity code with a word length n we have $X = \{0,1\}^n$, where X is the set of all binary words $x_1 \ldots x_n$ of n bits with $x_i \in \{0,1\}$, $n \geq 2$.

The subset ψ is the set of all binary words $x_1 \ldots x_n$ with $x_1 \oplus \ldots \oplus x_n = P$ with $P = 0$ for an even parity code and $P = 1$ for an odd parity code.

For $n = 3$, the set of all words of length 3 is $X = \{0,1\}^3 = \{000, 001, 010, 011, 100, 101, 110, 111\}$.

For $P = 0$, the even parity code ψ_{even} consists of $\psi_{even} = \{000, 011, 101, 101\}$. 011 is a code word since $011 \in \psi_{even}$. 111 is a non-code word since $111 \in X \backslash \psi_{even} = \{001, 010, 100, 111\} = \psi_{odd}$.

In this book only *block codes* are considered for error detection.

A code ψ is a block code of length n, $n \geq 1$, if ψ and X are sets of words of length n only. We assume $X = \{0,1\}^n$ and that the symbols or the components of the code words are binary.

Information is encoded and protected against certain errors by use of a code. If a code word is erroneously changed into a non-code word, the error will be detected.

If a correct code word $v = v_1, \ldots, v_n$ is erroneously changed into an erroneous word $v' = v'_1, \ldots, v'_n$ by an error, the error can be formally described by a binary error vector e.

The error vector e is defined as

$$e = (e_1, \ldots, e_n) = (v \oplus v') = (v_1 \oplus v'_1), \ldots, (v_n \oplus v'_n).$$

For $e_i = 1$ the ith component v_i of v is erroneously changed into $v'_i = v_i \oplus 1 = \overline{v}_i$, and for $e_j = 0$ the jth component $v'_j = v_j$ remains correct.

The number of ones in the error vector e indicates how many bit positions of v are erroneous.

The number $w(v)$ of ones in a binary vector $v = (v_1, \ldots, v_n)$,

$$w(v) = \sum_{i=1}^{n} v_i,$$

is called its *Hamming weight*.

The *Hamming distance* $d(v, v')$ between two vectors $v = v_1, \ldots, v_n$ and $v' = v'_1, \ldots, v'_n$ is defined as the number of different bit positions of these vectors v and v'. Since the Hamming distance $d(v, v')$ can be determined as

$$d(v, v') = \sum_{i=1}^{n} (v_i \oplus v_n),$$

we have

$$d(v, v') = w(v \oplus v'),$$

and the Hamming distance between two vectors v and v' is equal to the Hamming weight of the componentwise *XOR*-sum $(v \oplus v')$ of these vectors.

An error e with Hamming weight $w(e)$ will change $w(e)$ bits of the correct vector v.

For a given code C the Hamming distance between all pairs of its code vectors can be determined. The minimum distance d_{min} between all pairs of the code C is called the distance of the code.

If the distance of a code is d_{min}, no error e with a Hamming weight $w(e) \leq d_{min}$ can change a code word into another code word of that code, and all errors with a weight $w(e) \leq d_{min}$ will be detected by this code.

But we emphasize that not only errors e with a Hamming weight $w(e) \leq d_{min}$ are detected by this code.

Every error changing an expected code word into a non-code word is detectable, but an error changing an expected code word into another, unexpected code word is not.

If N_C is the number of code words, for every code word the number of non-detectable errors is $N_C - 1$. These are the errors changing the expected code word into another, unexpected code word.

In general, for two different code words the corresponding sets of detectable errors may be different. For linear codes these sets are equal and for non-linear codes these sets may be different. Therefore, *for linear codes* such as parity codes, group parity codes, Hamming codes or BCH-codes, independent of the expected code word, an error will be *with certainty* either detectable or not detectable. This is different for non-linear codes for which the detectability of an error depends on the expected code word. This will be demonstrated for the *Non-linear Split Error Detection Codes*, which will be introduced in the next subsection.

Very often the information which is to be encoded is given as an information word $u = u_1, \ldots, u_k$ of k bits. The information word is mapped to a code word $v = v_1, \ldots, v_n$ of n bits, $n > k$, by a function $g : \{0,1\}^k \rightarrow \{0,1\}^n$, $v = g(u)$.

Different information words u, u' with $u \neq u'$ have to be mapped to different code words v and v',

$$v = g(u) \neq g(u') = v'.$$

In many applications error detection circuits are designed by use of *systematic block codes*. (In some applications also *m-out-of-n codes* are used.)

We consider systematic block codes of length n and information words of length k, $k < n$.

In a code word of a systematic block code the information word $u = u_1, \ldots, u_k$ remains unchanged and $l = n - k$ check bits $c_1, \ldots, c_l$ are added to the k information bits to form a code word. Thus, for a code word v of a systematic block code we have

$$v = v_1, \ldots, v_n = g(u) = g(u_1, \ldots, u_k) = u_1, \ldots, u_k, c_1, \ldots, c_l \tag{3.7}$$

where

$$v_1, \ldots, v_k = u_1, \ldots, u_k. \tag{3.8}$$

The l check bits $c_1, \ldots, c_l$ are determined as

$$c_1 = g_1(u_1, \ldots, u_k), \ldots, c_l = g_l(u_1, \ldots, u_k) \tag{3.9}$$

and $g_1, \ldots, g_l$ are Boolean functions of length k.

If the Boolean functions $g_1, \ldots, g_l$ are all linear, the code is linear.

Now the classical error detection codes used in this book will be briefly described.

1. Parity Codes
   To the k information bits $u_1, \ldots, u_k$ a single check bit $c_1 = c_P$ (called the parity bit) is added, and a code word $v(u)$ of a parity code is of the form

$$v(u) = u_1 \ldots u_k c_P.$$

The parity bit c_P is defined as

$$c_P = u_1 \oplus \ldots \oplus u_k$$

for an even parity code and as

$$c_P = \overline{u_1 \oplus \ldots \oplus u_k}$$

for an odd parity code.

For an even parity code all the words of length $n = k+1$ with an even number of ones are code words, and all words of length $n = k+1$ with an odd number of ones are not.

For an odd parity code all the words of length $n = k+1$ with an odd number of ones are code words, and all words of length $n = k+1$ with an even number of ones are not.

2. Group Parity Codes
The information bits of a group parity code are partitioned into groups and for every group of information bits its own check bit, the parity bit of the group, is determined.

For l groups the corresponding l check bits $c_1, \ldots, c_l$ are determined by l parity equations

$$\begin{aligned} c_1 &= u_{1,1} \oplus \ldots \oplus u_{1,k_1}, \\ &\vdots \\ c_l &= u_{l,1} \oplus \ldots \oplus u_{l,k_l} \end{aligned}$$

with

$$u_{1,1}, \ldots, u_{1,k_1}, \ldots, u_{l,1} \ldots, u_{l,k_l} \in \{u_1, \ldots, u_k\}$$

and $k_1, \ldots, k_l \leq k$.

The groups $\{u_{1,1}, \ldots, u_{1,k_1}\}, \ldots, \{u_{l.1}, \ldots, u_{l,k_l}\}$ of information bits may be disjoint or not.

In general, every linear code is a group parity code. But the notion Group Parity Code is only used when the groups of information bits for which parity bits are computed are specially selected and adapted to the concrete circumstances.

For example, if the information bits are arranged as bytes, for every byte its own parity bit may be determined and the natural groups of information bits are the bytes.

If the input-output behaviour of a circuit is concurrently checked by a group parity code the circuit outputs are the information bits of the code. To achieve the best possible error detection probability for a given number of l parity bits, the circuit outputs are grouped into the best l groups by use of the net list of the circuit.

3. Duplication Code
Duplication (and comparison) is, as already described, the most frequently used method of concurrent checking. The information word $u = u_1 \dots u_k$ is simply duplicated. The code word $v(u)$ is

$$v(u) = u_1 \dots u_k \; c_1 \dots c_k$$

with

$$\begin{aligned} c_1 &= u_1, \\ &\vdots \\ c_k &= u_k. \end{aligned}$$

4. Two-Rail Code
The two-rail code is a simple modification of the duplication code. Instead of directly duplicating the bits of the information word $u = u_1 \dots u_k$ these bits are duplicated in inverse form. The code word $v(u)$ of the information word u is defined as

$$v(u) = u_1 \dots u_k \; c'_1 \dots c'_k$$

with

$$\begin{aligned} c'_1 &= \overline{u}_1, \\ &\vdots \\ c'_k &= \overline{u}_k. \end{aligned}$$

5. Berger Code
The check bits $c_1, \dots, c_l$ of a Berger code are the binary representation of the number of zeros (or the number of ones) of the corresponding information bits $u_1, \dots, u_k$.

Since, for k information bits, the number of zeros is at most k, the number l of the necessary check bits is

$$l = \lceil ld(k) + 1 \rceil.$$

$\lceil x \rceil$ denotes the largest integer less than or equal to x.

More formally, for the information word $u = u_1 \dots u_k$ the code word $v(u) = v_1 \dots v_n$ is

$$v(u) = v_1 \dots v_n = u_1 \dots u_k \; c_1 \dots c_l$$

with

$$c_1 \dots c_l = (\sum_{i=1}^{k} \overline{u}_i \,)_{binary},$$

where $(y)_{binary}$ is the binary representation of y.

For three information bits u_1, u_2, u_3 the $2 = \lceil ld(3) + 1 \rceil$ check bits c_1, c_2 are represented in Table 3.2.

Table 3.2 Check bits of a Berger code with 3 information bits

u_1	u_2	u_3	c_1	c_2
0	0	0	1	1
0	0	1	1	0
0	1	0	1	0
0	1	1	0	1
1	0	0	1	0
1	0	1	0	1
1	1	0	0	1
1	1	1	0	0

6. Modulo p Checking
   The check bits of Modulo p Checking are the arithmetic value of the data bits modulo p.
   The modulo p value r of an integer z,

$$r = z \text{ modulo } p,$$

is defined as the residue or the remainder r of the division of z by p,

$$z = m \cdot p + r, \quad \text{with} \quad 0 \leq r < p.$$

Thereby z, m, p and r are integers. Sometimes also negative residues with $0 \leq |r| < p$ are used.

Examples are:

$$\begin{aligned}
&7 \; mod \; 5 = 2, \\
&8 \; mod \; 3 = 5 \; mod \; 3 = 2, \\
&10 \; mod \; 2 = 0, \\
&2^n \; mod \; (2^n - 1) = 1, \\
&2^n \; mod \; (2^n + 1) = -1.
\end{aligned}$$

To the information bits $u_1, \ldots, u_k$ the check bits

$$c_1 \ldots c_l = [\, (u_k + 2u_{k-1} + 4u_{k-2} + \ldots + 2^{k-1}u_1) \; mod \; p \,]_{binary}$$

are added to form a code word $u_1 \ldots u_k c_1 \ldots c_l$.
$(u_k + 2u_{k-1} + 4u_{k-2} + \ldots + 2^{k-1}u_1)$ is the arithmetic value of $u_1 u_2 \ldots u_k$ and $[y]_{bin}$ denotes the binary representation of y.

For example, let $u_1u_2u_3u_4u_5 = 10111$ and $p = 3$. The arithmetic value of 10111 is

$$1 + 2 \cdot 1 + 4 \cdot 1 + 8 \cdot 0 + 16 \cdot 1 = 23.$$

Since 23 mod $3 = 2$ with $[2]_{binary} = 10 = c_1c_2$ the corresponding code word is

$$u_1u_2u_3u_4u_5 \; c_1c_2 = 10111 \; 10.$$

Since for $p \neq 2^n$ the determination of $z \ mod \ p$ for larger values of z can only be implemented with reasonable costs for $p = 2^n \pm 1$, in practice only these "low cost values" are used.

The reason is that the validity of the equations

$$2^{t \cdot s} mod \ (2^s - 1) = 1$$

and

$$2^{t \cdot s} mod \ (2^s + 1) = (-1)^t$$

allows a cascaded implementation of the modulo $p = 2^s \pm 1$ computation of the arithmetic value of the information bits [45]

For instance, for $p = 3 = 2^2 - 1$ we have

$$2^2 \ mod \ 3 = 2^4 \ mod \ 3 = 2^8 \ mod \ 3 = \ldots = 1,$$

and for $p = 5 = 2^2 + 1$

$$\begin{aligned} 2^{1 \cdot 2} \ mod \ 5 &= 2^{3 \cdot 2} \ mod \ 5 = \ldots = -1, \\ 2^{2 \cdot 2} \ mod \ 5 &= 2^{4 \cdot 2} \ mod \ 5 = \ldots = 1. \end{aligned}$$

By use of these equations

$$(32u_1 + 16u_2 + 8u_3 + 4u_4 + 2u_5 + u_6) \ mod \ 3$$

can be determined in a cascaded form as

$$(2u_1 + u_2) \ mod \ 3 \oplus_{mod \ 3} (2u_3 + u_4) \ mod \ 3 \oplus_{mod \ 3} (2u_5 + u_6) \ mod \ 3$$

and

$$(32u_1 + 16u_2 + 8u_3 + 4u_4 + 2u_5 + u_6) \ mod \ 5$$

as

$$(2u_1 + u_2) \ominus_{mod \ 5} (2u_3 + u_4) \oplus_{mod \ 5} (2u_5 + u_6)$$

with

$$x \ominus_{mod \ 5} \ (y \ mod \ 5) = x \oplus_{mod \ 5} \ (5 - [y_{mod \ 5}]).$$

Modulo p checking is especially useful for concurrent checking of arithmetic operations.

When checking arithmetic operations modulo p the following equations may be used

$$\begin{aligned} (x + y) \ mod \ p &= (x \ mod \ p) \oplus_{mod \ p} (y \ mod \ p), \\ (x - y) \ mod \ p &= (x \ mod \ p) \ominus_{mod \ p} (y \ mod \ p), \\ (x \cdot y) \ mod \ p &= (x \ mod \ p) \odot_{mod \ p} (y \ mod \ p). \end{aligned}$$

7. *m-out-of-n* Codes

The block length of an *m-out-of-n* code is n. Exactly m out of the n bits of the block are 1, and the remaining $n-m$ bits are 0. The number of code words of an *m-out-of-n* code is $\binom{n}{m}$.

Information bits and check bits cannot be separated.

For example, the $3 = \binom{3}{1}$ code words of an *1-out-of-3* code are 100, 010, 001 and the 5 remaining 3-bit words 000, 110, 101, 011, 111 are non-code words.

For the *2-out-of-4* code the $6 = \binom{6}{2}$ code words are 1100, 1010, 1001, 0110, 0101, 0011. All the $2^4 - \binom{4}{2} = 10$ remaining 4-bit words are non-code words.

3.2.2 Non-linear Split Error Detection Codes

In this subsection new non-linear systematic error detection codes are briefly described [46].

A *fixed number* of l check bits $c_1, \dots, c_l$ is added to the k, $k > 2$, information bits $u_1, \dots, u_k$. In general the number l of check bits is small. The expense for non-linear split error detection codes is between parity and Hamming codes.

At least one of the check bits, say c_1, is determined as

$$c_1 = f_{11}(u_{11}, u_{12}) \oplus f_{12}(u_{13}, u_{14}) \oplus \dots \tag{3.10}$$

where the set of variables $\{u_{11}, u_{12}, u_{13}, u_{14} \dots\}$ is a subset of the information bits $\{u_1, u_2, \dots, u_k\}$.

For $j = 1, 2, \dots$ the functions f_{1j} are two-input non-linear functions with a controlling input $con(1, j)$. Such functions are two-input *NAND-*, *AND-*, *NOR-* and *OR* functions with the controlling inputs 0, 0, 1 and 1 respectively.

A first example of a non-linear split error detection code with k information bits $u_1, u_2, \dots, u_k$, where k is even, and 3 check bits c_1, c_2, c_3 is shown in Fig. 3.10.

For this code the first check bit c_1 is determined as

$$c_1 = (\overline{u_1 \wedge u_2}) \oplus (\overline{u_3 \wedge u_4}) \oplus \dots \oplus (\overline{u_{k-1} \wedge u_k}) \tag{3.11}$$

where

$$\begin{aligned} f_{11}(u_{11}, u_{12}) &= (\overline{u_1 \wedge u_2}), \\ f_{12}(u_{13}, u_{14}) &= (\overline{u_3 \wedge u_4}), \\ &\vdots \, . \end{aligned}$$

All the functions f_{1j}, $j = 1, \dots, k/2$ are *NAND*-functions. The controlling value of the *NAND*-function is 0.

Similarly, the second check bit c_2 is determined as

$$\begin{aligned} c_2 &= f_{21}(u_{21}, u_{22}) \oplus f_{22}(u_{23}, u_{24}) \oplus \dots = \\ &= (\overline{u_1 \vee u_2}) \oplus (\overline{u_3 \vee u_4}) \oplus \dots \oplus (\overline{u_{k-1} \vee u_k}), \end{aligned} \tag{3.12}$$

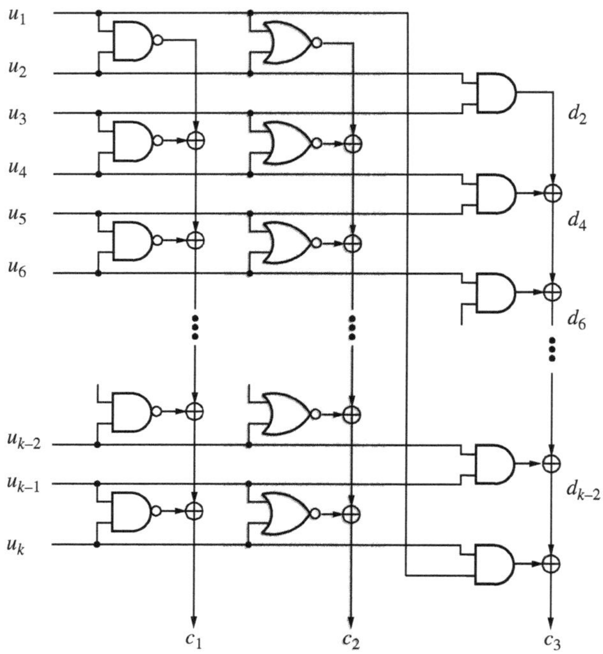

Fig. 3.10 Non-linear split error detection code with three check bits

where

$$\begin{aligned} f_{21}(u_{21}, u_{22}) &= (\overline{u_1 \vee u_2}), \\ f_{22}(u_{23}, u_{24}) &= (\overline{u_3 \vee u_4}), \\ &\vdots \end{aligned}$$

and all the functions f_{2j}, $j = 1, \ldots, k/2$ are *NOR*-functions. The controlling value for the *NOR*-function is 1.

Because of $(\overline{u_i \wedge u_j}) \oplus (\overline{u_i \vee u_j}) = u_i \oplus u_j$ we have

$$c_1 \oplus c_2 = u_1 \oplus u_2 \oplus u_3 \oplus u_4 \oplus \ldots \oplus u_{k-1} \oplus u_k = P(u), \tag{3.13}$$

and the linear parity function $P(u) = u_1 \oplus \ldots \oplus u_k$ of the information bits $u_1, \ldots, u_k$ may be considered as split into the two non-linear functions implemented by the check bits c_1 and c_2.

The third check bit c_3 is determined as

$$c_3 = (u_2 \wedge u_3) \oplus (u_4 \wedge u_5) \oplus \ldots \oplus (u_k \wedge u_1). \tag{3.14}$$

The error detection capability of the considered non-linear code will now be discussed and compared with the error detection capabilities of traditional error detection codes.

As has already been pointed out, for linear error detection codes such as parity code, Hamming code or BCH-code, the set of all possible errors E is divided into two disjoint subsets E_1 and E_2. Errors from the first subset E_1 are detected with certainty. Errors from the second E_2 will never be detected. Thus, for a parity code odd errors will always be detected and even errors will never be detected. This is different for a non-linear split error detection code. For these codes the set E of errors is divided into three disjoint subsets E'_1, E'_2 and E'_3.

Errors from the first subset E'_1 will always be detected by the non-linear split code as the errors from the first set E_1 of a linear error detection code. Errors from the second subset E'_2 will be detected at least with a probability of $1/2$. It is thereby assumed that all the information bits $u_1, u_2, \ldots, u_k$ are equal to 0 or to 1 with the same probability of $1/2$. Errors from the third subset E'_3, which is usually very small, will never be detected.

This fact will now be explained for the non-linear split error detection code with k information bits $u_1, \ldots, u_k$ and $l = 3$ check bits c_1, c_2, c_3 shown in Fig. 3.10.

In Fig. 3.10 the number of check bits is three. According to equation (3.13) the *XOR*-sum of the check bits c_1 and c_2 is equal to the parity $P(u)$ of the information bits $u_1, \ldots, u_k$. Therefore, every odd error of the information bits will always be detected either by the check bit c_1 or c_2 and all odd errors belong to E'_1.

All even errors for which not all the k information bits are simultaneously erroneous will be detected with at least a probability of $1/2$. They are elements of E'_2.

The error changing the information bits 1,0,1,0,..., 1,0... into 0,1,0,1,...,0,1,... cannot be detected by this code and belongs to E'_3.

We now explain how even-bit errors for which not all the k information bits $u_1, \ldots, u_k$ are simultaneously erroneous are detected. For these faults at least one of the check bits is changed with a probability greater or equal to $1/2$. We thereby assume, as already pointed, out that the information bits $u_1, \ldots, u_k$ are equal to 0 and 1 with a probability of $1/2$.

As an example of such an even-bit error we consider the 4-bit error for which the four correct information bits u_3, u_4, u_5, u_6 are changed into the erroneous information bits $\overline{u}_3$, $\overline{u}_4$, $\overline{u}_5$, $\overline{u}_6$. All the other information bits are assumed to be correct.

For the correct information bits we have, according to Fig. 3.10

$$
\begin{aligned}
d_2 &= u_2 \wedge u_3,\\
d_4 &= (u_2 \wedge u_3) \oplus (u_4 \wedge u_5),\\
d_6 &= (u_2 \wedge u_3) \oplus (u_4 \wedge u_5) \oplus (u_6 \wedge u_7)
\end{aligned}
$$

and for the erroneous inputs

$$
\begin{aligned}
d'_2 &= u_2 \wedge \overline{u}_3,\\
d'_4 &= (u_2 \wedge \overline{u}_3) \oplus (\overline{u}_4 \wedge \overline{u}_5),\\
d'_6 &= (u_2 \wedge \overline{u}_3) \oplus (\overline{u}_4 \wedge \overline{u}_5) \oplus (\overline{u}_6 \wedge u_7).
\end{aligned}
$$

For $d_6 \neq d'_6$ the considered 4-bit error will be detected since all the other values which are XORed with d_6 or d'_6 to form the check bit c_3 are the same for both d_6 and d'_6.

Let now $u_2 = 0$. For $d_6 \neq d'_6$ the error will be detected. If we assume $d'_6 = (\overline{u}_4 \wedge \overline{u}_5) \oplus (\overline{u}_6 \wedge u_7) = d_6 = (u_4 \wedge u_5) \oplus (u_6 \wedge u_7)$ the 4-bit error will not be detected by the check bit c_3. But then we have for $u_2 = 1$ with $(\overline{u}_4 \wedge \overline{u}_5) \oplus (\overline{u}_6 \wedge u_7) = (u_4 \wedge u_5) \oplus (u_6 \wedge u_7)$ that $d'_6 = \overline{u}_3 \oplus (\overline{u}_4 \wedge \overline{u}_5) \oplus (\overline{u}_6 \wedge u_7) \neq u_3 \oplus (u_4 \wedge u_5) \oplus (u_6 \wedge u_7) = d_6$, and the considered four-bit error will be detected for $u_2 = 1$. Similarly it can be shown that if the error will not be detected for input $u_2 = 1$, it will be detected for input $u_2 = 0$. Thus, the considered error will be detected for $u_2 = 0$ or for $u_2 = 1$ by c_3 and therefore at least with a probability of $1/2$.

We emphasize here that the input u_2 does not belong to the erroneous bits of the considered error. If the four-bit error changes $u_3 = 0$, $u_4 = 0$, $u_5 = 1$, $u_6 = 0$ into $\overline{u}_3 = 1$, $\overline{u}_4 = 1$, $\overline{u}_5 = 0$, $\overline{u}_6 = 1$, then this error will also be detected with certainty at the check bits c_1 and c_2 and this specific four-bit error will be detected with a probability of 1. An even-bit error with the first erroneous bit u_i, i odd, will be detected at the check bit c_3 with a probability of $1/2$. If i is even, the error will be detected at the check bits c_2 and c_3 with a probability of $1/2$.

Some of the even-bit errors for which all the information bits $u = u_1, \ldots, u_k$ are simultaneously erroneous cannot be detected by the code shown in Fig. 3.10.

An error for which all the information bits are erroneous cannot be detected if we have for $i = 1,2,3$

$$c_i(u_1, \ldots, u_k) = c_i(\overline{u}_1, \ldots, \overline{u}_k).$$

By use of

$$\begin{aligned} \overline{u_i \wedge u_j} &= 1 \oplus u_i \wedge u_j, \\ \overline{u_i \vee u_j} &= 1 \oplus u_i \oplus u_j \oplus u_i \wedge u_j, \end{aligned}$$

and

$$\overline{\overline{u}_i \vee \overline{u}_j} = u_i \wedge u_j$$

we obtain by direct calculation for $j = 1,2,3$

$$\begin{aligned} c_j(\overline{u}_1, \ldots, \overline{u}_k) &= c_j(u_1, \ldots, u_k) \oplus [(k/2)\ modulo2] \oplus u_1 \oplus u_2 \oplus \ldots \oplus u_k = \\ &= c_j(u) \oplus [(k/2)\ modulo2] \oplus P(u), \end{aligned} \tag{3.15}$$

and the error will not be detected if

$$[(k/2)\ modulo2] \oplus u_1 \oplus \ldots \oplus u_k = 0. \tag{3.16}$$

This is, for instance, the case if $u_1, \ldots, u_k = 1,0,1,0,\ldots$ is changed to $0,1,0,1,\ldots$ or if $u_1, \ldots, u_k = 1,1,1,1,\ldots$ for even $k/2$ is changed to $0,0,0,\ldots$.

The error will be detected if

$$[(k/2)\ modulo2] \oplus u_1 \oplus \ldots \oplus u_k = 1. \tag{3.17}$$

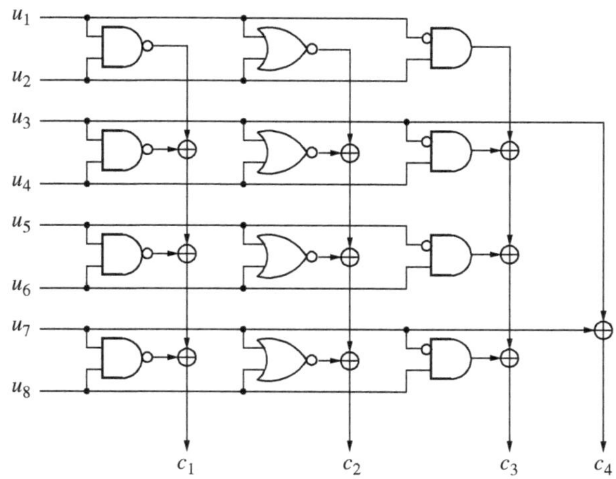

Fig. 3.11 Non-linear split error detection code with four check bits

In Fig. 3.11 for $k = 8$ another non-linear split error detection code with four check bits c_1, c_2, c_3, c_4 is shown. The number of check bits is again independent of the number of information bits. The three check bits c_1, c_2 and c_3 implement non-linear functions. The check bit c_4 is an XOR-sum of u_3 and u_7, and is a linear function.

Since we have

$$c_1 \oplus c_2 = u_1 \oplus u_2 \oplus \ldots \oplus u_7 \oplus u_8$$

all odd errors are detected. It can be shown that all error bursts of length 2, 4 and 6 are detected with certainty by this code. Many other errors are also detected with at least a probability of $1/2$.

Other examples of non-linear split error detection codes can be easily derived.

In this section basic notions and notations of systematic block codes were briefly explained. A new class of codes, non-linear split error detection codes, was introduced.

It was explained that parities of that codes are split into non-linear parts and it was demonstrated that these codes detect a first subset of errors with certainty, a second subset of errors with a probability greater or equal to $1/2$ and that a third, very small subset of errors will not be detected.

It was shown that the number of check bits is independent of the number of information bits and that the expenditure for the implementation for these codes is between parity codes and Hamming codes.

3.3 Parity and Group Parity Checking

Parity checking and group parity checking belong to the most popular methods of error detection. Conceptual simplicity combined with a moderate area overhead make this method attractive for designers. All odd errors at the circuit outputs or all odd errors within a group of outputs are detected. Even errors are not detectable.

Parity checking for regular structures such as adders, multipliers, dividers and others is different from parity checking for random logic. Regular structures are built up from some basic cells which can be specially designed to ease error detection by parity or group parity checking.

In this section we will only be interested in parity and group parity checking for random logic. The application of parity checking to regular structures will be detailed in the second part of this book where concurrent checking for different types of adders will be investigated.

In random logic during normal operation more than 80% of the errors which are caused by single stuck-at faults are 1-bit or odd-bit errors. These errors will be, as already pointed out, detected by parity checking. But for many applications 80+x% error detection is not high enough. Since the even-bit errors are not detected, efforts to improve parity and group parity checking are directed to reduce the number of two-bit and even-bit errors that simultaneously occur in a group of parity-checked outputs.

If the functional circuit to be checked is already designed and structurally given, then the circuit outputs have only to be divided into groups such that two-bit errors caused by single stuck-at faults do not simultaneously corrupt two outputs of the same group.

If the functional circuit is only functionally but not structurally specified, then the functional circuit can be designed in a first step as usual by an available synthesis tool and in a second step modified in such a way that errors caused by single stuck-at faults will only change single outputs of the parity-checked groups.

This section describes in detail parity and group parity checking as a special case of error detection using of systematic codes.

At the beginning predictor and generator circuits for systematic codes will be explained. It will be shown how a parity predictor can be derived by optimizing a serial connection of the functional circuit, which is given as a netlist of gates and an *XOR*-tree. The *XOR*-tree is in this way the corresponding generator circuit for parity checking.

It will be illustrated by a simple example how a fault in a single gate that is shared by two outputs can cause a two-bit error, which is not detectable by parity checking. If such a gate is duplicated, such a situation can be avoided.

To improve the error detection probability for parity and group parity checking, dependencies of circuit outputs with respect to gate faults will be considered in detail.

To systematically investigate such output dependencies the generalized circuit graph according to [7] for a combinational circuit will be described. This circuit graph will be extensively used in this section. The nodes of the circuit graph are the maximum sets of gates with one output. A first node and a second node are connected by an edge if the output of the first node is connected to an input of a gate belonging to the second node. It will be shown how this circuit graph can be easily derived from the netlist of the circuit.

The notions of independent and weakly independent outputs with respect to single gate faults will be explained.

Consequently, two outputs of a circuit are independent with respect to a fault if, in the presence of that fault, for an arbitrary input at most one of the considered outputs is erroneous.

Weakly independent outputs are introduced as a generalization of independent outputs.

Two outputs are weakly independent, if these outputs are either never erroneous or if there exists an input that for this input in the presence of the considered fault only one of the outputs is erroneous.

Structural and functional dependencies will be distinguished.

It will be shown how groups of independent outputs and groups of weakly independent outputs can be determined.

It will be demonstrated how groups of independent and weakly independent outputs can be used for the design of self-checking and self-testing circuits. Separate and joint implementations of the functional circuit and the predictor circuit will be considered as a result.

It will also be explained how the error detection probability can be improved by splitting special nodes of the generalized circuits graph, i.e. by replicating the gates belonging to the split nodes of the generalized circuit graph. Almost all the known methods for parity checking and group-parity checking will be explained in this section.

3.3.1 Predictor and Generator Circuits

For systematic block codes, error detection circuits are very often designed by use of predictor and generator circuits.

The general structure according to [35, 47] is shown in Fig. 3.12.

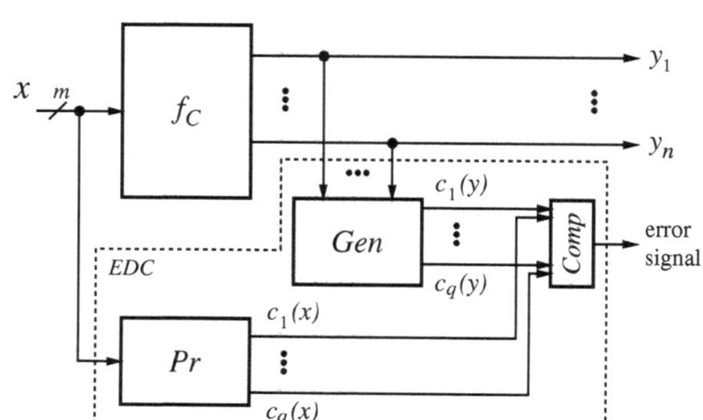

Fig. 3.12 Error detection by use of predictor and generator circuits

A combinational circuit f_C implementing the combinational function $y = f(x)$ with $y = y_1, \ldots, y_n$ and $x = x_1, \ldots, x_m$ is checked by an error detection circuit *EDC* consisting of a predictor *Pr*, a generator *Gen* and a comparator *Comp*.

The outputs $y_1, \ldots, y_n$ of the functional circuit f_C are determined as

$$y_1 = f_1(x),$$
$$\vdots$$
$$y_n = f_n(x),$$

where $f_1, \ldots, f_n$ are Boolean functions of length m.

If the combinational circuit f_C is monitored by use of a systematic code with n information bits $y_1, \ldots, y_n$ and q check bits $c_1, \ldots, c_q$, then we have for the check bits

$$\begin{aligned} c_1(y) &= g_1(y_1, \ldots, y_n), \\ c_2(y) &= g_2(y_1, \ldots, y_n), \\ &\vdots \\ c_q(y) &= g_q(y_1, \ldots, y_n), \end{aligned} \tag{3.18}$$

where $g_1, \ldots, g_q$ are Boolean functions of length n determined by the code. The generator *Gen* implements these Boolean functions and generates the check bits $c(y) = c_1(y), \ldots, c_q(y)$ depending on the outputs $y = y_1, \ldots, y_n$ of the circuit f_C.

Conversely, the predictor *Pr* generates the same check bits $c(x) = c_1(x), \ldots, c_q(x)$ but now depending on the inputs $x = x_1, \ldots, x_m$ of f_C according to

$$\begin{aligned} c_1(x) &= g_1\Big(y_1(x), \ldots, y_n(x)\Big) = g_1\Big(f_1(x), \ldots, f_n(x)\Big), \\ c_2(x) &= g_2\Big(y_1(x), \ldots, y_n(x)\Big) = g_2\Big(f_1(x), \ldots, f_n(x)\Big), \\ &\vdots \\ c_q(x) &= g_q\Big(y_1(x), \ldots, y_n(x)\Big) = g_q\Big(f_1(x), \ldots, f_n(x)\Big). \end{aligned} \tag{3.19}$$

The comparator *Comp* compares the check bits $c(x)$ with $c(y)$. If $c(y)$ and $c(x)$ are not equal an error signal *e* is generated.

Functionally, the predictor *Pr* is a serial connection of the monitored circuit f_C and the generator *Gen* as illustrated in Fig. 3.12. In practice, the predictor *Pr* is an optimized version of this serial connection of f_C and *Gen* obtained by use of an available synthesis tool as shown in Fig. 3.13.

In most cases one of the following approaches is used for the design of error detection circuits using of systematic codes:

1. The functional circuit f_C is given as a netlist of gates and this netlist will not be modified for the design of an error detection circuit. The expected errors at the circuit outputs of f_C are determined from the given netlist of f_C and the assumed

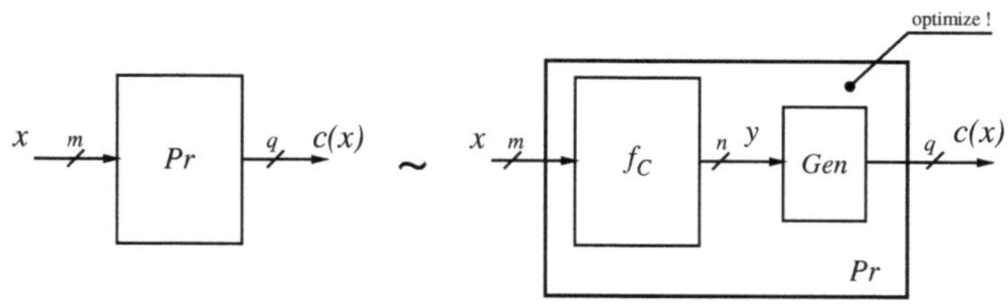

Fig. 3.13 Implementation of the predictor circuit for a systematic code

fault model. For these errors an error detection circuit with a high error detection probability is designed. In almost all cases the considered faults are single stuck-at faults.
2. The given netlist of the functional circuit f_C is modified in such a way that errors due to the considered faults of the modified circuit are relatively easily detectable by an error detection circuit. Again, in most cases errors due to single stuck-at faults are assumed. Unidirectional errors and errors with a limited multiplicity are also considered.
3. The design tool for the functional circuit f_C is modified in such a way that the necessary redundancy for error detection is automatically included in the structure of the functional circuit f_C. Again, single stuck-at faults are considered.

3.3.2 Parity Prediction

Error detection by parity prediction is illustrated in Fig. 3.14.

The functional circuit f_C has m binary inputs $x = x_1, \ldots, x_m$ and n binary outputs $y = y_1, \ldots, y_n$, $y_1 = f_1(x), \ldots, y_n = f_n(x)$.

The n outputs $y = y_1, \ldots, y_n$ of f_C are considered as the information bits of the parity code. To these n information bits a single check bit c_1, the parity bit, is added. This parity bit c_1 is determined twice, once as $c_1(y) = P(y)$ from the outputs y_1,

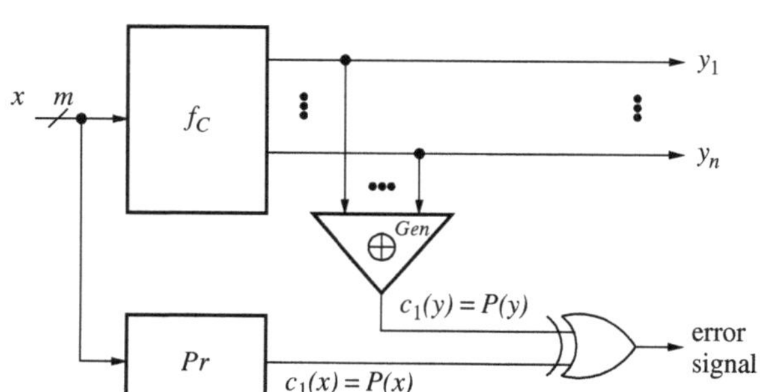

Fig. 3.14 Error detection by parity prediction

$y_2, \ldots, y_n$ of the functional circuit by the generator *Gen* and a second time as $c_1(x) = P(x)$ from the inputs $x_1, x_2, \ldots, x_m$ of the functional circuit by the predictor *Pr*.

The generator *Gen* determines the parity bit $P(y) = c_1(y)$ as

$$P(y) = c_1(y) = y_1 \oplus y_2 \oplus \ldots \oplus y_n \tag{3.20}$$

by an XOR-tree.

The predictor *Pr* computes $P(x) = c_1(x)$ as

$$P(x) = c_1(x) = f_1(x) \oplus f_2(x) \oplus \ldots \oplus f_n(x). \tag{3.21}$$

The predictor *Pr* implements an optimized form of the parity $P(x) = c_1(x)$ as shown in Fig. 3.15.

The comparator compares the one-dimensional outputs of the predictor *Pr* and the generator *Gen*. If a two-output comparator is used with $0,0$ and $1,1$ for correct output signals and $0,1$ and $1,0$ for erroneous output signals, no hardware for a comparator is needed.

For a single output comparator the outputs of the predictor *Pr* and of the generator *Gen* are to be XORed by a single XOR-gate.

All odd errors at the circuit outputs $y_1, \ldots, y_n$ of the functional circuit f_C are detected. But all even errors at this circuit outputs are not detected by parity prediction.

As a simple example we consider the functional circuit f_C presented in Fig. 3.16 in which we have

$$y_1(x) = (x_1 \wedge x_2) \oplus (x_2 \vee x_3),$$
$$y_2(x) = x_2 \vee x_3.$$

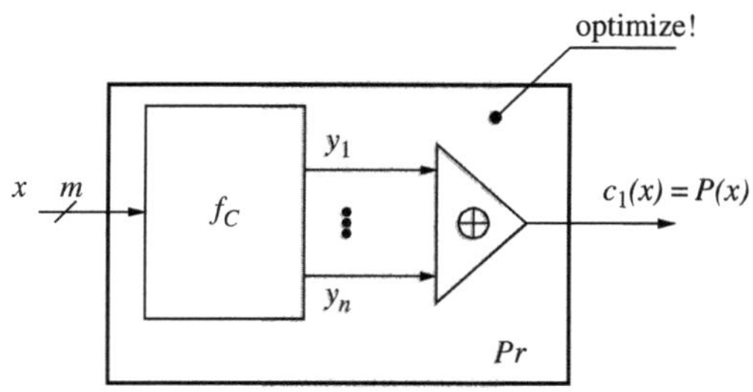

Fig. 3.15 Determination of parity predictor

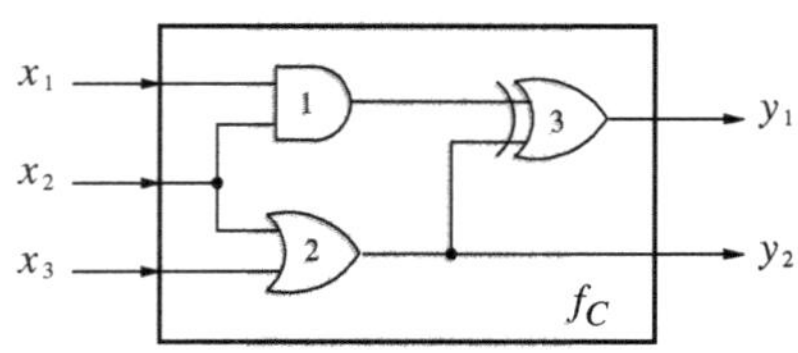

Fig. 3.16 Example of a combinational circuit

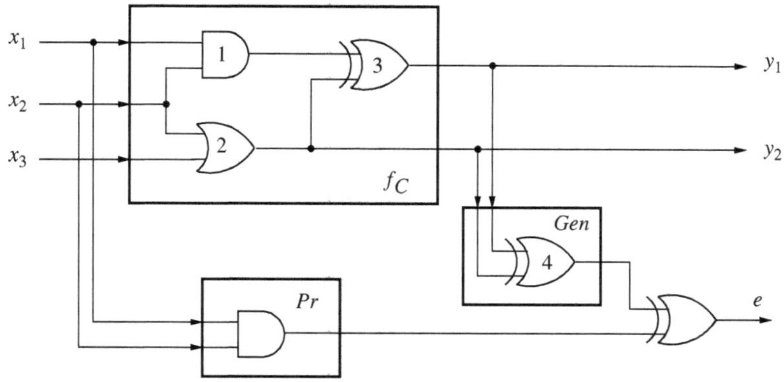

Fig. 3.17 Parity checking for the example of Fig. 3.16

Error detection by parity prediction is shown in Fig. 3.17 for this circuit.

The parity predictor *Pr* defines the parity $P(x)$ in relation to the inputs x_1, x_2, x_3 as given below:

$$P(x) = y_1(x) \oplus y_2(x) = x_1 \wedge x_2.$$

The generator *Gen* determines the parity of the outputs $P(y)$ as

$$P(y) = y_1 \oplus y_2.$$

The error signal e is obtained by XOR-ing $P(x)$ and $P(y)$,

$$e = P(y) \oplus P(x) = y_1 \oplus y_2 \oplus (x_1 \wedge x_2).$$

The OR-gate 2 in Fig. 3.17 is shared between the outputs y_1 and y_2. It is easy to see that no error at the output of that OR-gate can be detected by parity prediction, since the output of the OR-gate 2 is directly connected to the circuit output y_2 and via the XOR-gate 3 also with the circuit output y_1. Since the XOR-function is uniquely invertible every error at an input of the XOR-gate is always propagated to its output, which is also the circuit output y_1 and every error at the output of the OR-gate 2 always results in a two-bit error at the circuit outputs. But a two-bit error at the circuit outputs y_1 and y_2 is masked by the XOR-gate 4, which is the generator circuit in this simple example.

If the two circuit outputs of the functional circuit f_C are separately implemented without sharing common gates, all single gate faults, including single stuck-at faults, can be detected by parity checking. A separate implementation of the Boolean functions

$$\begin{aligned} y_1(x) &= (x_1 \wedge x_2) \oplus (x_1 \vee x_2), \\ y_2(x) &= x_2 \vee x_3 \end{aligned}$$

is shown in Fig. 3.18. Compared to Fig. 3.16 the OR-gate 2 is duplicated in the OR-gates 2 and 2′.

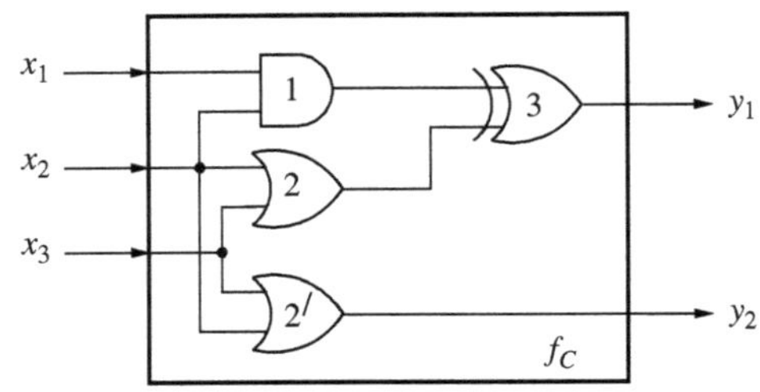

Fig. 3.18 Separate implementation for the outputs of the example in Fig. 3.16

The output of every gate is only connected to a single circuit output and every fault of a gate can at most erroneously change a single circuit output.

If the predictor is also separately implemented, every error at the circuit outputs due to a single gate fault will be immediately detected by parity checking.

3.3.3 Generalized Circuit Graph

To utilize dependencies between circuit outputs with respect to single gate faults in a systematic way the generalized circuit graph $G(f_C)$ of a combinational circuit f_C was introduced in 1970 in [7].

The circuit is supposed to be given as a netlist of gates. The nodes N_j, $j = 1, \ldots$ of the generalized circuit graph are the maximum sets of gates with a single output. Two nodes N_i, N_j are connected by an arc directed from N_i to N_j if the output of N_i is connected to an input of a gate belonging to the node N_j.

The generalized circuit graph $G(f_C)$ of a combinational circuit f_C can be determined from the netlist of f_C by the following algorithm.

Algorithm : Generalized Circuit Graph

The (non-fanout) outputs and the gates of the circuit are considered as elements

1. *Choose the elements which are the (non-fanout) outputs of the circuit f_C and mark these elements with different colors.*
2. *If an element (gate) is not yet marked and if its output is connected only to marked elements of the same color, then mark this element with the same color.*
3. *If the output of an element is not yet marked and if its output is connected only to already marked elements but with different colors, mark this element with a new color.*
4. *Continue until all elements are marked or colored.*

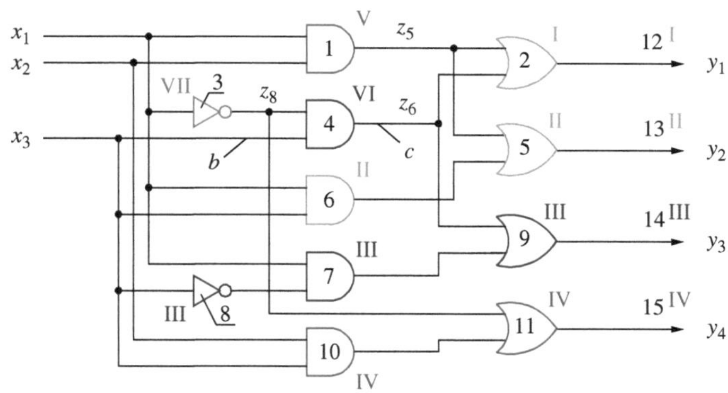

Fig. 3.19 Example of a combinational circuit

The elements with identical colors are the nodes of the generalized circuit graph $G(f_C)$.

As an example we now determine the generalized circuit graph for the circuit represented in Fig. 3.19, implementing at its four outputs the functions

$$
\begin{aligned}
y_1(x) &= (x_1 \wedge x_2) \vee (\bar{x}_1 \wedge x_3),\\
y_2(x) &= (x_1 \wedge x_2) \vee (x_1 \wedge x_3),\\
y_3(x) &= (x_1 \wedge \bar{x}_3) \vee (\bar{x}_1 \wedge x_3),\\
y_4(x) &= \bar{x}_1 \vee (x_2 \wedge x_3).
\end{aligned}
$$

The (non-fanout) outputs of the circuit are the outputs 12, 13, 14 and 15. They are marked with the colors I, II, III and IV.

Since the output of the element 2 is only connected to elements, here to the element 12, with color I, this element is also marked with color I.

Since the output of the element 5 is only connected to elements, with color II (element 13), the element 5 is also colored with color II. Similarly the element 6 is colored with color II.

The elements 14, 9, 7 and 8 are colored with color III, and the elements 15, 11 and 10 with color IV.

The output of the element 1 is connected with elements that are already differently colored. Therefore, the element is colored with a new color V. The elements 4 and 3 are also colored with new colors VI and VII, respectively.

The resulting generalized circuit graph is shown in Fig. 3.20.

The elements 12, 13, 14 and 15 corresponding to the circuit outputs y_1, y_2, y_3 and y_4 are underlined. The nodes I, II, III and IV containing the circuit outputs y_1, y_2, y_3 and y_4 are called the output nodes of these circuit outputs. We assign the variables z_5, z_6 and z_8 to the outputs of the (internal) nodes V, VI and VII .

From the construction of the generalized circuit graph it is evident that all faults in elements belonging to the same node as a circuit output can erroneously change only this circuit output. Faults in elements of a node which is connected to more

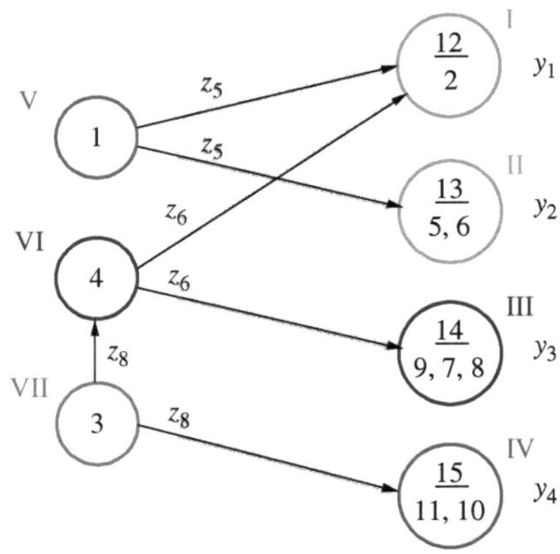

Fig. 3.20 Generalized circuit graph for the example of Fig. 3.19

than one node can possibly erroneously change more than one circuit output under some inputs.

In our example, a fault in gate 2 can erroneously change only the circuit output $12(=y_1)$. Faults in gate 1 can possibly simultaneously change the outputs $12(=y_1)$ and $13(=y_2)$ and faults in gate 4 can possibly result in a two-bit error at the outputs $12(=y_1)$ and $14(=y_3)$.

3.3.4 Independent Outputs and Weakly Independent Outputs

To express dependencies of outputs with respect to different faults the notions of *independent outputs* [17] and *weakly independent outputs* [18] are now given.

Definition 3.1. Independent Outputs:
Let f_C be a combinational circuit with the outputs $(y_1,\dots,y_n) = y$ and with the inputs
$(x_1,\dots,x_m) = x$ from the input set X and let $\Phi = \{\varphi_1,\dots\varphi_L\}$ the set of considered technical faults.

Then the outputs y_i and y_j are independent with respect to a fault φ, $\varphi \in \Phi$, if we have for all $x \in X$ either

$$y_i(x) = y_i(\varphi,x) \;\; and \;\; y_j(x) = y_j(\varphi,x)$$

or

$$y_i(x) \oplus y_j(x) \neq y_i(\varphi,x) \oplus y_j(\varphi,x).$$

Thereby $y_i(\varphi,x)$ denotes the output value at output y_i in the presence of the fault φ for input $x \in X$.

If the outputs y_i and y_j are independent with respect to a fault $\varphi \in \Phi$, then, in the presence of the fault φ, at most one of the outputs y_i or y_j is erroneous. Both these outputs are not erroneous at the same time.

Definition 3.2. The outputs y_i and y_j are independent with respect to a set of faults Φ if the outputs y_i and y_j are independent for all $\varphi, \varphi \in \Phi$.

Outputs may be *functionally* or *structurally* independent.

Definition 3.3. The outputs y_i and y_j are structurally independent if they are implemented without sharing any common gates.

If the outputs y_i and y_j are structurally independent, there is no node in the generalized circuit graph which is connected to both the outputs y_i and y_j respectively.

Structurally independent outputs are independent with respect to single stuck-at faults (and also with respect to any faults within the maximum classes of elements).

Definition 3.4. The outputs y_i and y_j are functionally independent if, in spite of their implementation with some common gates, the outputs y_i and y_j satisfy Definition 3.1.

The definition of independent outputs can be easily extended to a group of several outputs.

Definition 3.5. Let f_C be a combinational circuit with the outputs $y_1, \ldots, y_n$, the input set X and let φ be a technical fault.

Then the outputs $y_{i1}, \ldots, y_{iK}$ are a group of independent outputs of f_C with respect to a fault φ if we have for $x \in X$ either

$$\begin{aligned} y_{i1}(x) &= y_{i1}(\varphi, x), \\ &\vdots \\ y_{iK}(x) &= y_{iK}(\varphi, x) \end{aligned} \tag{3.22}$$

or

$$y_{i1}(x) \oplus y_{i2}(x) \oplus \ldots \oplus y_{iK}(x) \neq y_{i1}(\varphi, x) \oplus y_{i2}(\varphi, x) \oplus \ldots \oplus y_{iK}(\varphi, x). \tag{3.23}$$

The XOR-sum $y_{i1}(x) \oplus y_{i2}(x) \oplus \ldots \oplus y_{iK}(x)$ of the correct outputs of the considered group is different from the XOR-sum of the erroneous outputs $y_{i1}(\varphi, x) \oplus y_{i2}(\varphi, x) \oplus \ldots \oplus y_{iK}(\varphi, x)$ if one, three or an odd number of outputs are erroneous.

Theoretically, instead of Definition 3.5 a modified definition could be used where only a single output of a group is erroneous due to a fault.

Since the notion of *independent outputs* is mainly utilized for the design of parity-checked circuits with only a single parity bit or with some additional parity bits for some groups of outputs we prefer the definition of a group of independent outputs as given in Definition 3.5.

Similar to the Definitions 3.2 and 3.3 we have

Definition 3.6. The outputs $y_{i1}, \ldots, y_{iK}$ are a group of independent outputs of f_C with respect to a set of faults Φ if we have for $\varphi \in \Phi$ and for $x \in X$ either

$$y_{i1}(x) = y_{i1}(\varphi, x),$$
$$\vdots$$
$$y_{iK}(x) = y_{iK}(\varphi, x)$$

or

$$y_{i1}(x) \oplus y_{i2}(x) \oplus \ldots \oplus y_{iK}(x) \neq y_{i1}(\varphi, x) \oplus y_{i2}(\varphi, x) \oplus \ldots \oplus y_{iK}(\varphi, x).$$

Definition 3.7. The outputs $y_{i1}, \ldots, y_{iK}$ are a group of structurally independent outputs if all the outputs $y_{i1}, \ldots, y_{iK}$ are separately implemented without sharing any common gates.

A group of structurally independent outputs is independent with respect to all single gate faults, including single stuck-at faults.

In Fig. 3.19 the outputs $\{y_2, y_3\}$ and $\{y_2, y_4\}$ are groups of structurally independent outputs. This can be easily seen from the generalized circuit graph of Fig. 3.20. There is no common predecessor node for the output nodes II(y_2) and III(y_3). The same is true for the output nodes II(y_2) and IV(y_4).

Since in real designs independent outputs are seldom we now introduce the concept of *weakly independent outputs* according to [18]. Applying this concept, self-testing, but not self-checking circuits can be designed. Faults may be detected with some latency.

Definition 3.8. Weakly Independent Outputs:
Let f_C be a combinational circuit with the outputs $(y_1, \ldots, y_n) = y$ and with the inputs
$(x_1, \ldots, x_m) = x$ in a set X and let $\Phi = \{\varphi_1, \ldots \varphi_L\}$ be the set of considered technical faults.

Then the outputs y_i and y_j are *weakly independent* with respect to the set of faults Φ and with respect to a subset $\chi \subseteq X$ of inputs if for any $\varphi \in \Phi$ *there exists* an input $x \in \chi$ such that

$$y_i(x) \oplus y_j(x) \neq y_i(\varphi, x) \oplus y_j(\varphi, x) \tag{3.24}$$

or that for all $x \in \chi$

$$y_i(x) = y_i(\varphi, x) \quad and \quad y_j(x) = y_j(\varphi, x). \tag{3.25}$$

If the outputs y_i and y_j are weakly independent then there exists an input $x \in \chi$ such that in the presence of the fault φ one of the outputs is erroneous. For this input x both outputs are not erroneous at the same time.

As a modification of Definition 3.5 we have

Definition 3.9. Let f_C be a combinational circuit with the outputs $y_1, \ldots, y_n$ and with the inputs $x_1, \ldots, x_m = x$ in a set X and let $\Phi = \{\varphi_1, \ldots \varphi_L\}$ be the set of considered technical faults.

Then the outputs $y_{i1}, y_{i2}, \ldots, y_{iK}$ are a group of *weakly independent* outputs with respect to the set of faults Φ and with respect to a subset $\chi \subseteq X$ of inputs if for any $\varphi \in \Phi$ *there exists* an input $x \in \chi$ such that

$$y_{i1}(x) \oplus y_{i2}(x) \oplus \ldots \oplus y_{iK}(x) \neq y_{i1}(\varphi, x) \oplus y_{i2}(\varphi, x) \oplus \ldots \oplus y_{iK}(\varphi, x) \qquad (3.26)$$

or that for all $x \in \chi$, for all $\varphi \in \Phi$ and for $j = 1, \ldots, K$

$$y_{ij}(x) = y_{ij}(\varphi, x). \qquad (3.27)$$

The general structure of a self-checking or self-testing circuit designed by use of M groups of independent outputs or weakly independent outputs is shown in Fig. 3.21.

The combinational circuit f_C has n outputs $y_1, \ldots, y_n$. These outputs are divided into M (overlapping) groups of independent or weakly independent outputs. The parities $z_1, z_2, \ldots, z_M$ of the groups of independent or weakly independent outputs are determined by M parity trees $P_1, P_2, \ldots, P_M$. The inverted parities $\overline{z}_1, \overline{z}_2, \ldots, \overline{z}_M$ are realized by the additional predictor circuit g_C and compared with the parities z_1, $z_2, \ldots, z_M$ by the comparator *Comp*.

For weakly independent outputs we describe the proposed method in more detail.

We assume that $\chi \subseteq X$ is a test set for f_C.

Since the groups of outputs of f_C are weakly independent for any fault φ of the fault model there exists at least one input such that at least one of the parities $z_1, z_2, \ldots, z_M$ is erroneous in the test mode, and all faults will be detected in test mode.

In normal operation mode some of the errors due to the faults of the fault model are detected with latency. In any case, a fault is detected if the corresponding input

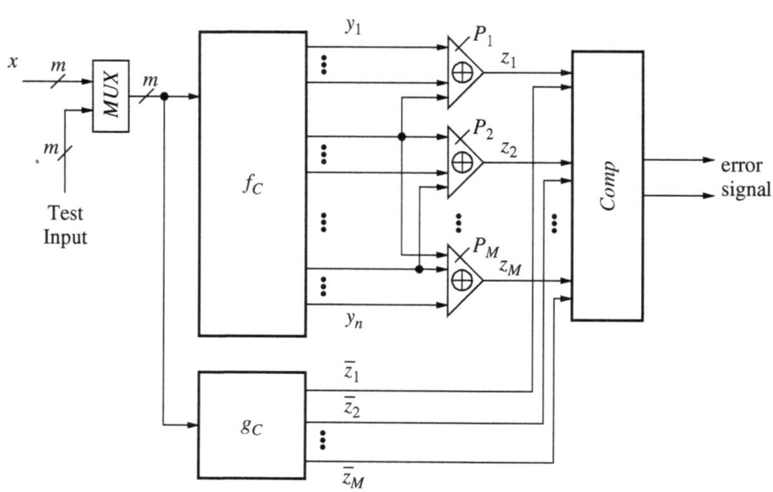

Fig. 3.21 Self-testing combinational circuit with weakly independent outputs

from the input set $\chi \subseteq X$ is applied to the circuit and in many cases also for other inputs from X.

3.3.5 Determination of Groups of Weakly Independent Outputs

The determination of *optimum* groups of weakly independent outputs is highly complex, and heuristic methods have to be used.

We consider a combinational circuit f_C with n outputs $y_1, \ldots, y_n$ and a set Φ of faults with a restricted input set χ. The restricted input set χ is supposed to be a test set for f_C.

The groups (or sets) of weakly independent outputs are denoted by $S_1, S_2, \ldots$.

As the first set S_1 of weakly independent outputs we choose the set of all outputs, $S_1 = y_1, \ldots, y_n$. These outputs are weakly independent for all faults $\varphi \in \Phi_{odd} \subseteq \Phi$ which change an odd number of circuit outputs at least for some input x from the restricted input set χ, $x \in \chi$.

Now we remove Φ_{odd} from the set of all faults Φ and we obtain Φ_{even} as

$$\Phi_{even} = \Phi \setminus \Phi_{odd}.$$

Practically speaking, the set Φ_{even} consists of 0% to 17% of all faults [48, 49].

The set Φ_{even} consists of all faults for which there is no input from the test set χ with an odd number of erroneous outputs.

To every fault $\varphi_k \in \Phi_{even} = \{\varphi_1, \ldots, \varphi_L\}$ we assign an input $x_k \in \chi$ such that at least one output y_j of f_C is erroneous

$$y_j(x_k) \neq y_j(\varphi_k, x_k). \tag{3.28}$$

Simultaneously with the output y_j for input x_k at least a second output of f_C is erroneous.

We define an (L, n) matrix $T_{k,l}$. The rows of $T_{k,l}$ correspond to the faults of Φ_{even} and the columns to the different outputs of f_C. The matrix $T_{k,l}$ is specified as

$$T_{k,l} = \begin{cases} 1 \text{ if } y_l(x_k) \neq y_l(\varphi_k, x_k) \\ 0 \text{ if } y_l(x_k) = y_l(\varphi_k, x_k). \end{cases} \tag{3.29}$$

If $T_{k,l} = 1$ the lth output of f_C is erroneous for input x_k due to the fault $\varphi_k \in \Phi_{even}$.

For every column l, $l = 1, \ldots, n$ we count the number $|1_l|$ of ones in that column. $|1_l|$ indicates how often the output y_l is erroneous for all the faults $\varphi_k \in \Phi_{even}$ for the inputs x_k that are assigned to the faults φ_k.

Any zero-column m corresponding to an output y_m with no error will be removed from the matrix $T_{k,l}$. The corresponding output y_m will never be erroneous due to faults $\varphi_k \in \Phi_{even}$ for the input x_k.

Now we describe a heuristic algorithm for the determination of groups of weakly independent outputs.

Algorithm : Determination of Groups of Weakly Independent Outputs

1. *From the unmarked columns of the matrix $T_{k,l}$ we select a column r with a maximum value $|1_r|$ of ones and we mark this column with *.*
2. *All the rows j for which $T_{j,r} = 1$ are also marked with *. (These rows correspond to the faults φ_j, $\varphi_j \in \Phi_{even}$ for which for input x_j the output $y_r(x_j)$ is erroneous, i.e. for which $y_r(x_j) \neq y_r(\varphi_j, x_j)$. These faults are detected by observing the output y_r.)*
3. *Now all the columns m, $m \neq r$ for which $T_{j,m} = 1$ are marked with +. (If m' is a column marked with +, then we have $T_{j,m'} = 1$ and $T_{j,r} = 1$ and the outputs $y_{m'}$ and y_r are simultaneously erroneous if the fault φ_j occurs. Therefore the outputs $y_{m'}$ and y_r should not be added modulo 2 and they should not be in the same group of weakly independent outputs.*
   In general, outputs corresponding to columns marked by + should not be in the same group of independent outputs.)
4. *If not all columns are marked either by + or * goto 1, otherwise goto 5.*
5. *End.*

The columns marked by * correspond to a group of weakly independent outputs.

Next the columns marked by * are deleted from the matrix $T_{k,l}$. Every zero-column will also be removed and the next group of weakly independent outputs is determined. The procedure is repeated until all rows are removed from the matrix.

The groups of weakly independent outputs consist of the first group S_1 of all outputs and the following mutually disjoint subsets $S_2, S_3, \ldots, S_M$ of the outputs which are determined by the described algorithm.

Since $S_2, S_3, \ldots, S_M$ are mutually disjoint, the parity z_1 of the outputs from S_1, i.e. the parity of all circuit outputs, can be determined as the XOR-sum of the parities $z_2, z_3, \ldots, z_m$ of the outputs of these groups $G_2, G_3, \ldots, G_M$.

The parities $z_1, z_2, \ldots, z_M$ of the groups $G_1, G_2, \ldots, G_M$ of weakly independent outputs are compared with the corresponding inverted parities $\bar{z}_1, \bar{z}_2, \ldots, \bar{z}_M$ determined by the predictor circuit g_C and compared by the comparator *Comp*.

We now illustrate the method described for the example represented in Fig. 3.22.

The circuit of Fig. 3.22 with three inputs x_1, x_2 and x_3 and three outputs y_1, y_2 and y_3 implements at its outputs the following Boolean functions

$$
\begin{aligned}
y_1(x_1,x_2,x_3) &= (x_1 \wedge x_2) \vee (\bar{x}_1 \wedge x_3),\\
y_2(x_1,x_2,x_3) &= (x_1 \wedge x_2) \vee (x_1 \wedge x_3),\\
y_3(x_1,x_2,x_3) &= (x_1 \wedge \bar{x}_3) \vee (\bar{x}_1 \wedge x_3).
\end{aligned}
$$

The set $S_1 = \{y_1, y_2, y_3\}$ is chosen as the first group S_1 of weakly independent outputs. These outputs are XORed to form the parity bit z_1.

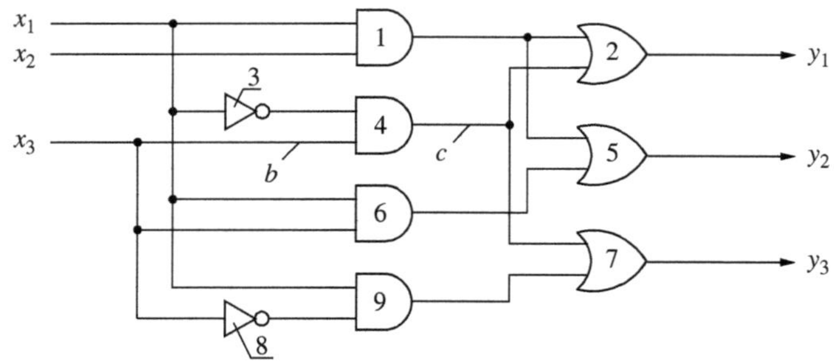

Fig. 3.22 Example of a combinational circuit

It can be easily proven by fault simulation that the group of outputs $S_1 = \{y_1, y_2, y_3\}$ is weakly independent with respect to all single stuck-at faults except the stuck-at-1 fault $b/1$ at the input b of gate 4 and the stuck-at-0 fault $c/0$ at the output c of gate 4. The set Φ_{even} is $\Phi_{even} = \{b/1, c/0\}$.

In the case of the stuck-at fault $b/1$ the circuit implements

$$\begin{aligned} y_1(b/1, x_1, x_2, x_3) &= (x_1 \wedge x_2) \vee \overline{x}_1, \\ y_2(b/1, x_1, x_2, x_3) &= (x_1 \wedge x_2) \vee (x_1 \wedge x_3), \\ y_3(b/1, x_1, x_2, x_3) &= (x_1 \wedge \overline{x}_3) \vee \overline{x}_1. \end{aligned}$$

In the presence of the stuck-at fault $c/0$ we have

$$\begin{aligned} y_1(c/0, x_1, x_2, x_3) &= x_1 \wedge x_2, \\ y_2(c/0, x_1, x_2, x_3) &= (x_1 \wedge x_2) \vee (x_1 \wedge x_3), \\ y_3(c/0, x_1, x_2, x_3) &= x_1 \wedge \overline{x}_3. \end{aligned}$$

We assign the inputs $x_{b/1} = (0,0,0)$ and $x_{c/0} = (0,0,1)$ to the faults $y(b/1)$ and $y(c/0)$ respectively.

For the input $(0,0,0)$ we have

$$1 = y_1(b/1; 0,0,0) \neq y_1(0,0,0) = 0,$$
$$0 = y_2(b/1; 0,0,0) = y_2(0,0,0) = 0$$

and

$$1 = y_3(b/1; 0,0,0) \neq y_3(0,0,0) = 0$$

and for input $(0,0,1)$

$$0 = y_1(c/0; 0,0,1) \neq y_1(0,0,1) = 1,$$
$$0 = y_2(c/0; 0,0,1) = y_2(0,0,1) = 0$$

and

$$0 = y_3(c/0; 0,0,1) \neq y_3(0,0,1) = 1.$$

The corresponding (2×3) matrix T is

$$T = \begin{pmatrix} 1\,0\,1 \\ 1\,0\,1 \end{pmatrix}.$$

The matrix $T_{k,l}$ has two rows corresponding to the two faults $b/1$ and $c/0$ from Φ_{even}. The assigned inputs are 000 and 001 respectively.

The matrix $T_{k,l}$ has three columns 1,2,3 corresponding to the three circuit outputs y_1, y_2, y_3.

The matrix element T_{11} is $T_{11} = 1$ since for the first fault $b/1$ the output $y_1(b/1;0,0,0)$ is different from the correct output $y_1(0,0,0)$. Because of $y_2(b/1;0,0,0) = y_2(0,0,0)$ and $y_3(b/1;0,0,0) \neq y_3(0,0,0)$ we have $T_{12} = 0$ and $T_{13} = 1$.

Similarly the elements $T_{21} = 1$, $T_{22} = 0$ and $T_{23} = 1$ are determined since $y_1(c/0;0,0,1) \neq y_1(0,0,1)$, $y_2(c/0;0,0,1) = y_1(0,0,1)$ and $y_3(c/0;0,0,1) \neq y_3(0,0,1)$.

The outputs y_1 and y_3 are erroneous for two faults, the output y_2 for no faults. Therefore we have $|1_1| = |1_3| = 2$ and $|1_2| = 0$. These numbers are shown in Table 3.3 in the row "sum".

The steps of the algorithm for the determination of the remaining groups of weakly independent outputs are illustrated in the following Tables 3.3 and 3.4.

Column 2 with no error will be removed from T, and the new matrix is given in Table 3.4.

One of the columns of the new matrix with a maximum sum of errors is column 2 corresponding to the output y_3. This column is marked with *. Since $T_{12} = T_{22} = 1$ the rows 1 and 2 will also be marked by *. These rows correspond to faults $b/1$ and

Table 3.3 Determination of groups of weakly independent outputs: step 1

input	*fault*	*outputs*		
		y_1	y_2	y_3
000	$b/1$	**1**	**0**	**1**
001	$c/0$	**1**	**0**	**1**
sum		2	0	2

Table 3.4 Determination of groups of weakly independent outputs: step 2

input	*fault*	*outputs*	
		$y_1 +$	$y_3 *$
000	$b/1*$	**1**	**1**
001	$c/0*$	**1**	**1**
sum		2	2

$c/0$ for which the output y_3 is erroneous for the assigned inputs $0,0,0$ and $0,0,1$. These faults are detected by observing the output y_3.

Because of $T_{11} = 1$ the column 1, corresponding to the output y_1 has to be marked with +, and the output y_1 should not be *XOR*ed with output y_3, since for the fault $b/1$ the outputs y_3 and y_1 are simultaneously erroneous under input $0,0,0$.

Now the rows 1 and 2 which are marked by * will be deleted from T and the resulting matrix is empty.

The groups of weakly independent outputs are $S_1 = \{y_1, y_2, y_3\}$ and $S_2 = \{y_3\}$ and the predictor has to inplement

$$\overline{y_1(x) \oplus y_2(x) \oplus y_3(x)} = \overline{x}_1 \vee (x_2 \wedge x_3)$$

and

$$\overline{y}_3 = (x_1 \wedge x_3) \vee \overline{x}_1 \vee (\overline{x}_3).$$

The resulting *self-testing* circuit is shown in Fig. 3.23.

3.3.6 Circuit Modification by Node-Splitting

If circuit outputs of a given circuit f_C are not weakly independent with respect to single stuck-at faults the circuit f_C can be modified such that the outputs of the modified circuit are weakly independent.

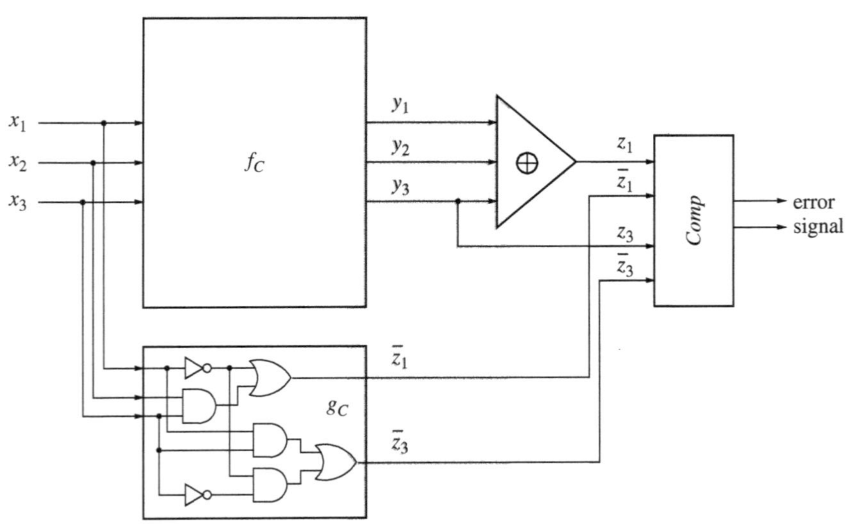

Fig. 3.23 Self-testing combinational circuit with weakly independent outputs for the example of Fig. 3.22

The gates belonging to specific nodes are s times replicated if the fan-out of the node is s.

This method was introduced in [50] and it is called *node-splitting*.

To explain the method we consider a circuit for which the functional outputs and the (inverted) parity are jointly optimized [51].

By node splitting we are able to achieve that the circuit becomes self-testing with respect to all single stuck-at faults. Most of the errors, due to single stuck-at faults are immediately detected.

As an example we consider the circuit of Fig. 3.19 which implements at its output y_4 the Boolean function

$$y_4 = \overline{y_1(x) \oplus y_2(x) \oplus y_3(x)} = \bar{x}_1 \vee (x_2 \wedge x_3),$$

which is the inverted parity of the outputs y_1, y_2 and y_3.

The group of four outputs $\{y_1, y_2, y_3, y_4\}$ is weakly independent with respect to all single stuck-at faults except the single stuck-at-1 fault $b/1$ and the single stuck-at-0 fault $c/0$ of gate 4.

In the generalized circuit graph of Fig. 3.20 the node VI consists of the gate 4 only. Since this node has a fan-out of 2 the node, i.e. gate 4, has to be duplicated. The modified circuit is shown in Fig. 3.24.

Functionally the circuits of Figs. 3.19 and 3.24 are identical. Structurally they are different since in the circuit of Fig. 3.24 the gate 4 of Fig. 3.19 is duplicated in the gates 4 and 4′.

The generalized circuit graph of the modified circuit is given in Fig. 3.25.

For the modified circuit it can be shown that the group of outputs $\{y_1, y_2, y_3, y_4\}$ is weakly independent with respect to all single stuck-at faults of the gates 1 and 3. For all the other gates it is evident from the generalized circuit graph that the group of

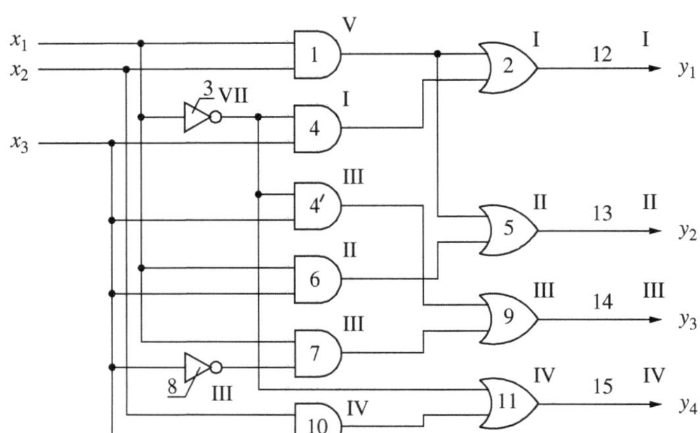

Fig. 3.24 Modified by node-splitting circuit of Fig. 3.19

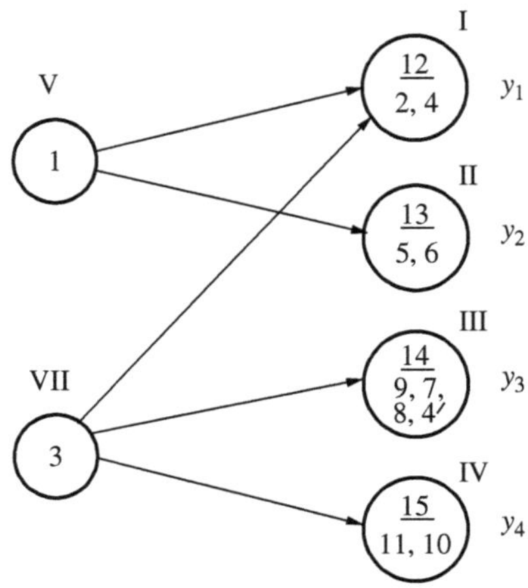

Fig. 3.25 Generalized circuit graph of the modified circuit in Fig. 3.24

outputs $\{y_1, y_2, y_3, y_4\}$ is even independent with respect to all single stuck-at faults of these gates.

The outputs y_1, y_2, y_3 are *XOR*ed and compared with the output y_4 which implements the (predicted) inverted parity $\overline{P}(x)$ as shown in Fig. 3.26.

In this approach the parity and the functional outputs of the original circuit can be jointly implemented. The circuit modification by *node-splitting* guarantees that the group of all outputs is weakly independent with respect to all single stuck-at faults and that the resulting circuit is *self-testing*.

If it is possible to modify the design tools, the method proposed in [52] is of interest.

The method is based on the dependencies of the nodes of the generalized circuit graph and is applicable to arbitrary combinational circuits with not too many outputs.

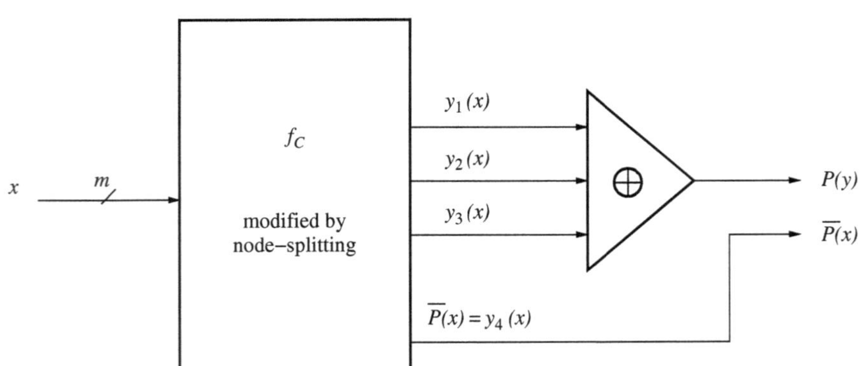

Fig. 3.26 Self-testing parity checking with joint implementation of the functional bits and the parity bit

For every node of the generalized circuit graph of the functional combinational circuit the set of structurally reachable outputs is determined. These sets are described by reachability vectors. The reachability vectors determine the structurally possible errors at circuit outputs due to faults of elements of the corresponding nodes.

The original circuit is assumed to have only reachability vectors corresponding to errors detectable by the selected group parity code.

The automatic synthesis tool is modified in such a way that *resubstitution* and *extraction of subexpressions* can only be executed if the resulting possible structural errors corresponding to the obtained reachability vectors are still detectable by the considered group parity code.

Starting with a single parity group for every output, i.e. with duplication and comparison, a heuristic algorithm is developed step by step to enlarge the number of outputs in the parity groups until the functional circuit, the predictor circuit and the necessary comparator or two-rail checker are at optimum.

It is of interest to notice that using the proposed method for a functionally given circuit under check the *optimal group parity code*, the structure of the functional circuit and the predictor are automatically determined.

3.3.7 Further Methods for the Determination of Weakly Independent Outputs

There are several other methods to determine groups of weakly independent outputs. We mention the following:

- Based on path lengths from the output of the gates to the circuit outputs, groups of weakly independent outputs are determined in [53].
- In [54] in a first step groups of outputs which depend upon disjoint input sets are determined. These are groups of independent outputs.
  Depending on the fault coverage the remaining outputs are either added to already existing groups or they form new groups of outputs.
  If all the remaining outputs form their own group of one output only, then all the groups are groups of independent outputs and the fault coverage for single stuck-at faults is 100%.
- In [55] first a test set is computed by an ATPG generator. Then all the outputs of the considered circuit are XORed to form a parity tree.
  The faults that are not detected at the output of the parity tree by the inputs from the test set are determined and additional circuit outputs are added until 100% fault coverage for single stuck-at faults (in test mode) is achieved.
- A probabilistic method for the determination of groups of weakly independent outputs is given in [56].
  Based on simple estimates for the existence of sensitized paths from the internal signal lines to the circuit outputs, the circuit outputs are partitioned into groups

of weakly independent outputs. No fault simulation is needed. Since all the used algorithms are of linear complexity with respect to the number of gates and of quadratic complexity with respect to the number of circuit outputs the method can be applied to large circuits.

- The application of a Hamming Code for the determination of output groups is described for instance in [57]. The concrete circuit structure is not taken into consideration, and the self-testing property cannot be guaranteed. However, for many circuits most of the errors at the circuit outputs due to single stuck-at faults are 1-bit and 2-bit errors. Since 1-bit and 2-bit errors are not masked by a Hamming code the fault coverage is high.

Parity and group parity checking was described in this section. It was shown how a predictor circuit and a generator circuit have to be added to the functional circuit. In this way, the generator circuit is an *XOR*-tree to implement the *XOR*-function of the outputs of the functional circuit.

If the functional circuit is given as a netlist of gates, it was shown how the parity predictor can be easily derived by optimizing a serial connection of the functional circuit and an *XOR*-tree which is the generator circuit already described.

It was illustrated by a simple example how a fault in a single gate, which is shared by two outputs, can cause a two-bit error that is not detectable by parity checking. If such a gate is duplicated, a situation of this kind can be avoided.

To improve the error detection probability for parity and group parity checking, dependencies of circuit outputs with respect to gate faults were considered.

It was demonstrated that output dependencies can be systematically investigated by use of the generalized circuit graph according to [7]. The determination and the application of this circuit graph for the design of self-checking and self-testing circuits were described in detail.

The notions of independent and weakly independent outputs with respect to single gate faults were given.

It was explained that two outputs of a circuit are independent with respect to a fault if, in the presence of that fault, for an arbitrary input at most one of the considered outputs is erroneous.

Weakly independent outputs were defined as a generalization of independent outputs.

Two outputs are weakly independent, if they are either never erroneous or if there exists an input such that for this input in the presence of a fault only one of the outputs is erroneous.

Structural and functional dependencies were distinguished.

It was shown how groups of independent outputs and groups of weakly independent outputs can be determined and how these groups of independent and weakly independent outputs can be utilized for the design of self-checking and self-testing circuits.

It was also explained how the error detection probability can be improved by splitting special nodes of the generalized circuits graph, i.e. by replicating the gates belonging to the split nodes of the generalized circuit graph.

Separate and joint implementations of the functional and predictor circuits were considered.

Consequently, a separate implementation results in a better error detection probability combined with a higher necessary area overhead, but for a joint implementation the error detection probability is only slightly reduced, especially if some specific nodes of the generalized circuit graph are split.

3.4 Odd and Even Error Detection

In this section it will be shown how, according to [58], odd and even errors can be detected by two different error detection circuits, a first error detection circuit for all the odd errors and a second one for the even errors caused by single stuck-at faults.

The odd errors are, as usual, detected by parity prediction.

It will be explained how the even error detection circuit can be determined by fault simulation for single stuck-at faults. For small circuits, all the input values are simulated for every fault.

It will be stated how a truth table can be derived for a partially defined Boolean function which defines the even error detection circuit.

Heuristic solutions with a high probability of even error detection can be applied for larger circuits.

3.4.1 Description of Odd and Even Error Detection

In [58] it is proposed to detect odd and even errors by two different error detection circuits. Since the method is also applicable to larger circuits it will be described in some detail.

Most of the errors due to single stuck-at faults are single-bit errors. These single-bit errors and also the other odd errors are detected by ordinary parity checking, and the predicted parity is determined by the parity predictor. The few even errors resulting from single stuck-at faults are detected by an additional *Even-Bit Error Detection Circuit*.

In Fig. 3.27 the odd and even error detection for a combinational circuit f_C with an m-dimensional input $x = (x_1, \ldots, x_m)$ and an n-dimensional output $y = (y_1, \ldots, y_n)$ is shown.

The design of the parity predictor *Pr* was already explained in Section 3.3.2 and will not be repeated here.

We now describe how the *Even Error Detection Circuit* can be designed. The set of all considered single stuck-at faults is denoted by Φ.

We explain the method for small circuits by use of the truth table for the *Even Error Detection Circuit*.

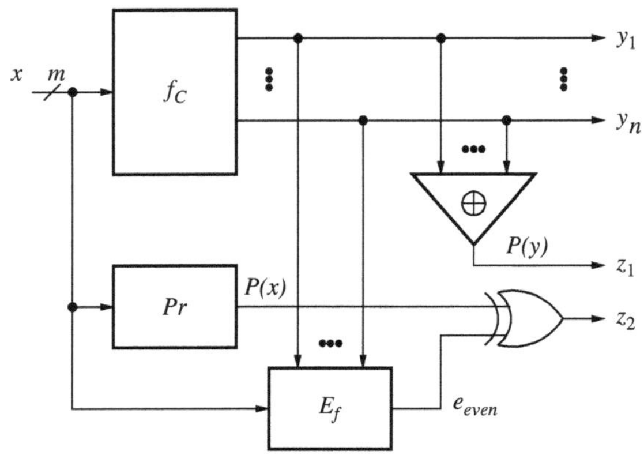

Fig. 3.27 Odd-even error detection

The truth table of the error detection function $E_f(x,y)$, $x \in \{0,1\}^m$, $y \in \{0,1\}^n$ of the *Even Error Detection Circuit* can be determined in the following steps:

1. Inject a fault $\varphi \in \Phi$ into f_C and remove φ from Φ, $\Phi = \Phi \setminus \{\varphi\}$.
2. Apply all inputs $x \in \{0,1\}^m$ to $f_{C,\varphi}$, where $f_{C,\varphi}$ denotes the combinational circuit with the injected fault φ.
   If for some x the output $y(\varphi,x) = y_1(\varphi,x),\dots,y_n(\varphi,x)$ differs from $y(x) = y_1(x),\dots,y_n(x)$ in an even number of bits, add $x, y(\varphi,x)$ to the on-set of the truth table of the error detection function E_f.
3. If Φ is empty, go to 4, otherwise go to 1.
4. Add for $x \in \{0,1\}^m$ $x, y(x)$, $y(x) = f(x)$, to the off-set of the truth table of the error detection function E_f.
5. All the rows not yet defined of the truth table for E_f are "don't care".
6. End.

The *Even Error Detection Circuit* is an implementation of the truth table of the function $e_{even}(x) = E_f(x,y)$. The output e_{even} of the *Even Error Detection Circuit* can be XORed with the output of the parity predictor.

Error detection of odd and even errors at the outputs z_1 and z_2 is summarized in the following Table 3.5.

Table 3.5 Error detection of odd and even errors

	z_1	z_2
no error	0	0
	1	1
odd error	0	1
	1	0
even error	0	1
	1	0

If *no error* occurs, we have $P(x) = P(y)$, $e_{even} = 0$ and therefore $z_1 = z_2$. In the case of an *odd error* we obtain $P(x) \neq P(y)$, $e_{even} = 0$ and $z_1 \neq z_2$. For an *even error* we have $P(x) = P(y)$, $e_{even} = 1$ and also $z_1 \neq z_2$, and odd and even errors due to single stuck-at faults are always detected.

In reality the set Φ of all single stuck-at faults can be reduced. If from the location of a fault φ there are only paths to a single circuit output, this fault can be removed from Φ since due to this fault no even-bit error can be caused. This reduction can be easily done by use of the generalized circuit graph as described in Section 3.3.3.

For larger circuits the on-set of the even error detection function E_f can be approximated as described in [58]. The aim of the method proposed in [58] is to compact groups of input values of the truth table for E_f by input cubes with "don't cares".

For every fault $\varphi \in \Phi$ randomly selected inputs $x \in \{0,1\}^n$ are applied to $f_{C,\varphi}$. If for a randomly selected input $x = x_1,\ldots,x_m$ the corresponding output $y(\varphi,x) = y_1(\varphi;x_1,\ldots,x_i,\ldots,x_m)\ldots,y_n(\varphi;x_1,\ldots,x_i,\ldots,x_m)$ has an even number of erroneous bits, then all the bits $x_1,\ldots,x_m$ of the input x are inverted one by one.

For $i = 1,\ldots,m$ the input $x_1,\ldots,\bar{x}_i,\ldots,x_m$ is applied to $f_{C,\varphi}$.

If $y_1(\varphi;x_1,\ldots,\bar{x}_i,\ldots,x_m)\ldots,y_n(\varphi;x_1,\ldots,\bar{x}_i,\ldots,x_m)$ also has an even number of erroneous bits, then the ith bit x_i in the input vector x is replaced by an undefined value (don't care) and the corresponding input cube belongs to the on-set of E_f. If no even-bit error occurs, the bit $\bar{x}_i$ is flipped back, and the unmodified input vector is added to the on-set of E_f.

In this method several input vectors with "don't care" values, or several input cubes, are generated. Only the input cubes, all their input vectors resulting in even-bit errors are fault-simulated for the determination of the on-set of the error detection function E_f. Details are described in [58].

In this section it was explained how odd and even errors can be detected by two different error detection circuits.

It was detailed that the odd errors are detected, as usual, by parity prediction while the even errors, caused by single stuck-at faults, are detected by an additional even error detection circuit.

It was explained how the even error detection circuit can be determined as a partially defined combinational circuit by fault simulation.

3.5 Code-Disjoint Circuits

Code-disjoint circuits are needed to detect input errors (and not only errors caused by internal faults) during normal operation.

The inputs and the outputs of a code-disjoint circuit have to be encoded. Input code words are mapped to output code words and input non-code words to output non-code words.

All the errors changing an input code word into a non-code word are detected.

Code-disjoint circuits for parity-encoded inputs and outputs will be described in this section.

It will be shown how an arbitrarily given combinational circuit can be systematically modified into a code-disjoint circuit. The inputs and outputs have to be parity-encoded and, compared to ordinary parity prediction, basically the predicted parity has to be replaced by the *XOR*-sum of the input parity and the predicted output parity.

Two different types of code-disjoint circuits, the first one with two additional outputs and the second one with three outputs will be introduced in this section. These code-disjoint circuits are especially suitable for error detection of different serial connections of code-disjoint circuits.

Code-disjoint partial duplication will also be explained in this section. Using code-disjoint partial duplication, soft errors directly induced in the registers of the circuit, errors due to transient faults in the combinational part and input errors are detected.

The method of code-disjoint partial duplication is widely applied for error detection for adders as described in Chapter 4 and for multipliers [8, 9] and dividers [10].

3.5.1 Design of Code-Disjoint Circuits

Now we describe how for an arbitrarily given combinational circuit f_C a code-disjoint circuit $f_{C,cd}$ for parity codes can be systematically designed. The inputs and the outputs of the code-disjoint circuit are parity-encoded. Input errors and output errors due to internal faults are detected. The content of the section is based on [59].

Let f_C be an arbitrarily given combinational circuit with m inputs $(x_1,\ldots,x_m) = x$ and n outputs $(y_1,\ldots,y_n) = y$ implementing the Boolean Functions $y_1(x) = f_1(x),\ldots, y_n(x) = f_n(x)$ for which the code-disjoint circuit has to be designed.

To the functional inputs $(x_1,\ldots,x_m)$ an additional parity input x_{m+1}, where

$$x_{m+1} = P(x) = x_1 \oplus x_2 \oplus \ldots \oplus x_m$$

is added. $P(x)$ is called the input parity.

We also add two auxiliary additional outputs y'_{n+1} and y'_{n+2} with

$$\begin{aligned} y'_{n+1} &= P(x) \oplus P\Big(y(x)\Big) = \\ &= x_1 \oplus x_2 \oplus \ldots \oplus x_n \oplus y_1(x) \oplus y_2(x) \oplus \ldots \oplus y_n(x) \end{aligned}$$

and

$$y'_{n+2} = x_{m+1} = P(x).$$

The output y'_{n+2} is directly connected with the input parity x_{m+1}. The output y'_{n+1} implements the *XOR*-sum of the input parity $P(x)$ and the parity $P\Big(y(x)\Big)$ of the outputs of f_C in accordance with x.

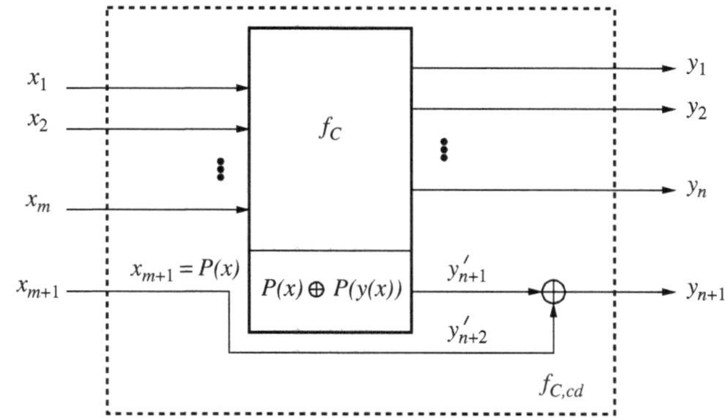

Fig. 3.28 Code-disjoint combinational circuit with a single additional output

The outputs y'_{n+1} and y'_{n+2} are XORed to form the output y_{n+1}, where

$$y_{n+1} = y_1(x) \oplus y_2(x) \oplus \ldots \oplus y_n(x) = P\Big(y(x)\Big).$$

The resulting circuit in Fig. 3.28 is the code-disjoint circuit $f_{C,cd}$.

If the circuit f_C is given as a netlist of gates, the code-disjoint circuit $f_{C,cd}$ can be obtained by optimizing the circuitry of Fig. 3.29 by an available synthesis tool.

The circuits implementing $P(x) \oplus P(y)$ and f_C may be either jointly or separately implemented.

More sophisticated implementations are described in [60].

We show now that the circuit $f_{C,cd}$ of Fig. 3.28 is code-disjoint with respect to a parity code.

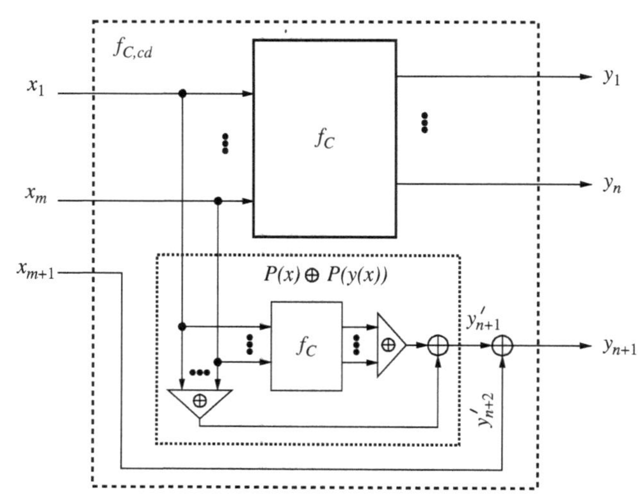

Fig. 3.29 Synthesis of a code-disjoint circuit from a netlist of the functional circuit

By construction $f_{C,cd}$ maps parity-encoded inputs $x_1,\ldots,x_m,x_{m+1}$ with an odd number of ones into parity-encoded outputs $y_1,\ldots,y_n,y_{n+1}$, also with an odd number of ones.

If an erroneous input $x_1^e,\ldots,x_m^e,x_{m+1}^e$ with an even number of ones is applied to $f_{C,cd}$ and if the number of ones in the corresponding output $y_1^e,\ldots,y_n^e,y_{n+1}^e$ is even, the error will be detected.

Let us now assume that the number of ones in $x_1^e,\ldots,x_m^e,x_{m+1}^e$ is even and that in the corresponding outputs $y_1^e,\ldots,y_n^e,y_{n+1}^e$ the number of ones is odd. Then we change x_{m+1}^e into $\overline{x}_{m+1}^e$ and y_{n+1}^e is changed into $\overline{y}_{n+1}^e$.

Now the number of ones in $x_1^e,\ldots,x_m^e,\overline{x}_{m+1}^e$ is odd and the number of ones in $y_1^e,\ldots,y_n^e,\overline{y}_{n+1}^e$ is even.

This is a contradiction since inputs of odd parity are mapped by $f_{C,cd}$ to outputs of odd parity, and we have shown that inputs of even parity are mapped to outputs of even parity. Thus the circuit $f_{C,cd}$ is indeed code-disjoint.

A serial connection of three code-disjoint circuits $f_{C,cd}^1$, $f_{C,cd}^2$, $f_{C,cd}^3$ is presented in Fig. 3.30.

Only the outputs of the third code-disjoint circuit have to be checked by comparing the corresponding parity bits $r_1 = P(y^3)$ and $r_2 = y_{n_3+1}^3$.

As a modification of the code-disjoint circuit with a single additional output $y_{n+1} = P(x) \oplus P(y) \oplus x_{m+1}$ a code-disjoint circuit with three additional outputs $y'_{n+1} = P(y(x))$, $y'_{n+2} = P(x)$ and $y_{n+3} = x_{m+1}$ can be designed, as shown in Fig. 3.31.

In Fig. 3.31 the additional output y'_{n+1} implements the parity of the outputs in relation to x,

$$y'_{n+1} = y_1(x) \oplus y_2(x) \oplus \ldots \oplus y_n(x) = P\Big(y(x)\Big).$$

For the additional output y'_{n+2} we have

$$y'_{n+2} = x_1 \oplus x_2 \oplus \ldots \oplus x_m = P(x).$$

and the additional output y_{n+3} is directly connected to the parity input x_{m+1}.

According to [60] internal nodes of f_C can be utilized for the implementation of y'_{n+2}.

By comparing y_{n+3} and x_{m+1} input errors and errors due to some internal faults of f_C are detected. Other internal faults of f_C are detected by comparing $y'_{n+1} = y_1(x) \oplus y_2(x) \oplus \ldots \oplus y_n(x) = P(y(x))$ with the XOR-sum $P(y) = y_1 \oplus \ldots \oplus y_n$ of the outputs of f_C.

A serial connection of three code-disjoint circuits $f_{C,cd}^1$, $f_{C,cd}^2$, $f_{C,cd}^3$ with three additional outputs is shown in Fig. 3.32.

The input parity $P(x^1)$ generated at the output $y_{n_1+2}^{1'}$ of $f_{C,cd}^1$ is compared with $x_{m_1+1}^1$.

The output parity $P(y^1)$ at the output $y_{n_1+1}^{1'}$ of $f_{C,cd}^1$ is equal to the input parity $P(x^2)$ at the output $y_{n_2+2}^{2'}$ of the circuit $f_{C,cd}^2$, and the output parity $P(y^2)$ at the output $y_{n_2+1}^{2'}$ of $f_{C,cd}^2$ is equal to the input parity $P(x^3)$ at the output $y_{n_3+2}^{3'}$ of the circuit $f_{C,cd}^3$.

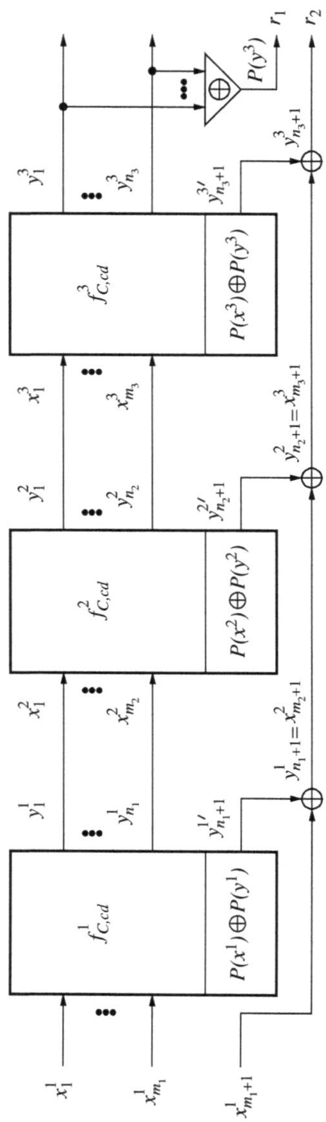

Fig. 3.30 Serial connection of code-disjoint circuits

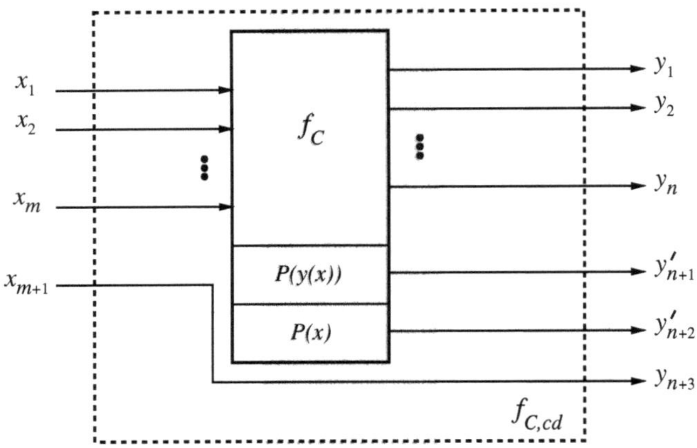

Fig. 3.31 Code-disjoint circuit with three additional outputs

The output parity $P(y^3)$ of $f^3_{C,cd}$ at the output $y^{3'}_{n_3+1}$ is compared with the XOR-sum $y^3_1 \oplus \ldots \oplus y^3_{n_3}$ of the outputs $y^3_1, \ldots, y^3_{n_3}$ of $f^3_{C,cd}$.

Now we describe how a code-disjoint circuit can be designed for a partially duplicated functional circuit as shown in Fig. 3.33.

The original circuit f_C with m inputs $X = x_1, \ldots, x_m$ and the n outputs implementing the n Boolean functions of length m $z_1 = f_1(x), \ldots, z_n = f_n(x)$ is implemented as a serial connection of the circuits f_{C_1} and f_{C_2}. The circuit f_{C_1} has the inputs $x = x_1, \ldots, x_m$ and the outputs $y = y_1, \ldots, y_p$ which are also the inputs of f_{C_2}. The outputs $z = z_1, \ldots, z_n$ of f_{C_2} are also the outputs of the original circuit f_C. The circuit f_{C_2} is duplicated in $f^1_{C_2}$ with the outputs $z^1_1, \ldots, z^1_n$, and $f^2_{C_2}$ with the outputs $z^2_1, \ldots, z^2_n$. The outputs of these duplicated circuits are stored in the two registers R^1 and R^2.

The inputs $x = x_1, \ldots, x_m$ are parity-encoded with the parity bit

$$x_{m+1} = x_1 \oplus \ldots \oplus x_m = P(x).$$

At the line y_{p+1} the XOR-sum $P(x) \oplus P(y(x))$ of the input parity $P(x)$ and the output parity $P(y(x))$ of the outputs $y_1, \ldots, y_p$ of f_{C_1}

$$P(x) \oplus P\Big(y(x)\Big) = x_1 \oplus \ldots \oplus x_m \oplus y_1(x) \oplus \ldots y_p(x)$$

are determined as a function of x.

The XOR-sum of the outputs $y_1, \ldots, y_p$ of f_{C1} is determined by an XOR-tree and XORed with $P(x) \oplus P(y(x))$ to form the input parity $P(x)$, which is compared with x_{m+1}.

All odd input errors and all odd errors at the outputs of f_{C1} are detected by comparing x_{m+1} and $P(x) = y_{p+1} \oplus y_1 \oplus \ldots \oplus y_p$.

All errors due to faults in one of the subcircuits $f^1_{C_2}$ or $f^2_{C_2}$ or in one of the two registers R^1 or R^2 are detected by comparing the contents of the registers R^1 and R^2.

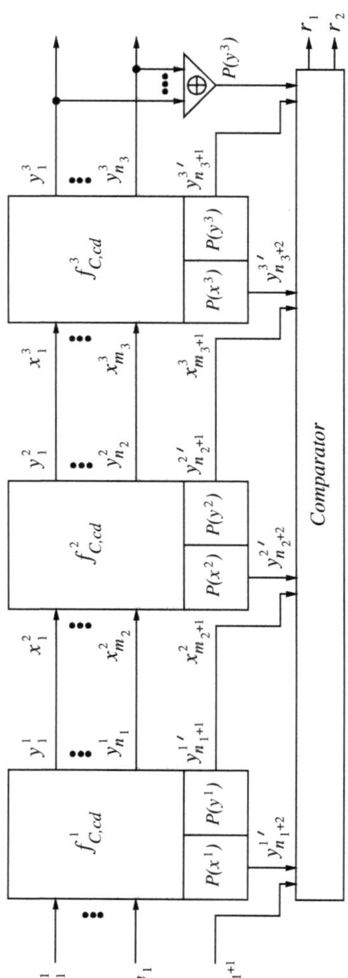

Fig. 3.32 Serial connection of code-disjoint circuits with three additional outputs

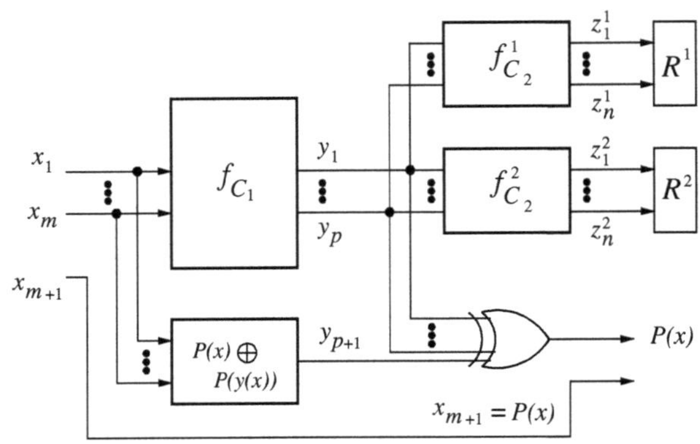

Fig. 3.33 Partially duplicated code-disjoint circuit

In this section it was shown how code-disjoint circuits for parity-encoded inputs and parity-encoded outputs can be systematically designed for a combinational circuit, which is given as a netlist of gates.

It was explained that basically the predicted output parity of ordinary parity checking has to be replaced by the *XOR*-sum of the input parity and the predicted output parity where this *XOR*-sum can be either jointly or separately optimized with the functional circuit.

It was demonstrated how for serial connections code-disjoint circuits with either two or three additional outputs can be implemented.

To also detect input errors, errors due to transient faults in the combinational part and soft errors directly induced by radiation in the output registers the concept of code-disjoint partially duplicated circuits was explained.

3.6 Error Detection by Complementary Circuits

This section presents the method of error detection by complementary circuits.

It will be shown how for an arbitrarily given functional circuit such a complementary circuit can be determined such that the componentwise *XOR*-sum of the outputs of the functional and complementary circuits will be a code word of a chosen code, as long as no error occurs. Errors in the functional or complementary circuit are detected by a code checker if the componentwise *XOR*-sum of the outputs of the functional and complementary circuits is a non-code word. The number of inputs of the checker is only equal to the number of outputs of the functional circuit, and no additional control bits have to be added.

Almost arbitrary codes such as *m-out-of-n* codes, Berger codes, systematic and non-systematic linear codes can be used and the error detection can be easily adapted to the codes that already used in the system.

It will be explained that a complementary circuit for a completely defined functional circuit is only "partially specified" since for every input the *XOR*-sum of the outputs of the functional circuit and of the complementary circuit has not to be a specific but an arbitrary code word. If the number of code words of the considered code is N then for every input the corresponding output of the functional circuit can be complemented by N different possible outputs of the complementary circuits. For all these possible N outputs of the complementary circuit the *XOR*-sum with the output of the functional circuit is one of the N code words of the code. One of these N possible outputs has to be selected to specify the complementary circuit such that the complementary circuit has a small area. It remains a challenging problem to utilize this plurality of possible specification of the complementary circuit for a systematic circuit optimization.

This is different to the traditional method of error detection by use of systematic codes. In this method, where only systematic codes and, for instance, no *m-out-of-n* code can be used, the predictor and generator circuits are completely specified by the given functional circuit and the chosen systematic code, and in this case the design of an optimal error detection circuit is reduced to the standard CAD problem to optimize the completely specified generator and predictor circuits in Fig. 3.12.

It will be explained in detail how the method of error detection by complementary circuits can be applied for a *1-out-of-n* code.

It will be explained how the circuit outputs can be divided into groups of three outputs and how for these groups the corresponding complementary outputs are determined. Thereby different heuristics for the determination of the complementary circuit will be discussed where the functional circuits are supposed to be given as a netlist of gates.

Until now error detection by complementary circuits is the only method of error detection for which necessary and sufficient conditions for the existence of totally self-checking checkers can be proven.

This proof will be given in this section for a *1-out-of-n* code. Also very simple sufficient conditions will be derived.

It will be shown that these conditions are very weak and that they are fulfilled for almost all circuits.

The new method of error detection by complementary circuits is a method with great potential which needs further investigation.

3.6.1 Error Detection by Use of Complementary Circuits

Error detection by use of complementary circuits was introduced in [61, 62].

This method is shown in Fig. 3.34.

For a given functional circuit f_C a complementary circuit g_C is determined in such way that the componentwise XOR-sum of the corresponding outputs of the functional circuit f_C and of the complementary circuit g_C is an element or a code

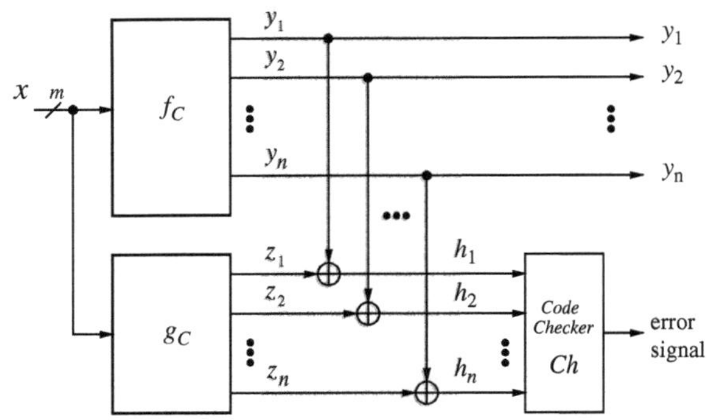

Fig. 3.34 Error detection by a complementary circuit

word of a chosen code as long as no error occurs. The code checker *Ch* checks the code word property.

The functional circuit f_C has m binary inputs $x = (x_1, \ldots, x_m) \in \{0,1\}^m$ and n outputs $y = (y_1, \ldots, y_n)$. f_C implements the n Boolean functions

$$y_1(x) = f_1(x),$$
$$\vdots$$
$$y_n(x) = f_n(x).$$

The complementary circuit g_C has the same m binary inputs $x = (x_1, \ldots, x_m) \in \{0,1\}^m$ and g_C has the n outputs $z = (z_1, \ldots, z_n)$. g_C implements at its outputs the n Boolean functions

$$z_1(x) = g_1(x),$$
$$\vdots$$
$$z_n(x) = g_n(x).$$

Consequently, the complementary circuit g_C has to be designed in such a way that for all inputs $x \in X$ the XOR-sum

$$h_1(x) = f_1(x) \oplus g_1(s),$$
$$\vdots$$
$$h_n(x) = f_n(x) \oplus g_n(x)$$

of the outputs of the functional circuit f_C and of the complementary circuit g_C is an element or code word of the considered code *Cod*.

Every block code of length n can be used as a code. Examples of possible codes are *m-out-of-n* codes with $n > m > 0$, arbitrary linear block codes or Berger codes of block length n.

We emphasize that the componentwise XOR-sum of the outputs of f_C and g_C must not be a specific code word of the considered code but an arbitrary code word. Therefore, even if the considered code *Cod* is already selected, the complementary circuit g_C is only "partially specified".

This will now be explained in more detail.

Let $Cod = \{C^1, \ldots, C^N\}$ be the considered code of N code words

$$\begin{aligned} C^1 &= (c_1^1, \ldots, c_n^1), \\ &\vdots \\ C^N &= (c_1^N, \ldots, c_n^N) \end{aligned}$$

of length n. Then the Boolean functions $g_1(x), \ldots, g_n(x)$ have to be determined such that for $x \in X$

$$\begin{aligned} h_1(x) &= f_1(x) \oplus g_1(x), \\ &\vdots \\ h_n(x) &= f_n(x) \oplus g_n(x) \end{aligned}$$

for $h(x) = \big(h_1(x), \ldots, h_n(x)\big) \in Cod = \{C^1, \ldots, C^N\}$.

Although the Boolean functions $f_1(x), \ldots, f_n(x)$ implemented by the functional circuit f_C are completely specified for all $x \in X$ we have for every $x \in X$ N different choices for the output values $g_1(x), \ldots, g_n(x)$ of g_C to form one of the N code words of *Cod*.

With $X = \{0,1\}^m$ there are N^{2^m} functionally different complementary circuits g_C, and until now it was a challenging problem to utilize these huge number of possibilities for an optimum design of a complementary circuit g_C for a given combinational circuit f_C.

(Remember that the situation is completely different for the design of an error detection circuit by use of a predictor and a generator circuit. The predictor *Pr* and the generator *Gen* are functionally completely specified by the check bits of the chosen systematic code. The determination of the predictor and the generator circuit is reduced to the optimization of a full functionally specified combinational circuit by a CAD tool.)

3.6.2 Complementary Circuits for 1-out-of-3 Codes

Next we describe how the method of error detection by complementary circuits can be practically applied for a *1-out-of-3-code* [63, 64].

First we explain the method for a functional circuit f_C with three outputs y_1, y_2 and y_3 only.

At the outputs y_1, y_2 and y_3 the circuit f_C implements the Boolean functions

$$y_1(x) = f_1(x),$$
$$y_2(x) = f_2(x)$$
and
$$y_3(x) = f_3(x).$$

We suppose that the functional circuit f_C is given as a netlist of gates and we show in Fig. 3.35 how a complementary circuit g_C^{no} that has not yet been optimized can be derived from f_C.

g_C^{no} implements at its outputs z_1, z_2 and z_3 the Boolean functions

$$z_1(x) = g_1(x) = 0,$$
$$z_2(x) = g_2(x) = \Big(f_1(x) \wedge f_2(x)\Big) \vee \Big(\overline{f_1(x) \vee f_2(x) \vee f_3(x)}\Big),$$
and
$$z_3(x) = g_3(x) = \Big(f_1(x) \vee f_2(x)\Big) \wedge f_3(x).$$

The complementary circuit g_C is obtained by optimizing the circuit g_C^{no} by an available synthesis tool.

Concurrent checking by use of the complementary circuit g_C is presented in Fig. 3.36.

The inputs h_1, h_2 and h_3 of the *1-out-of-3* checker are determined as

$$h_1(x) = y_1(x) \oplus z_1(x) = f_1(x) \oplus 0 = f_1(x),$$
$$h_2(x) = y_2(x) \oplus z_2(x) =$$
$$= f_2(x) \oplus \bigg(\Big(f_1(x) \wedge f_2(x)\Big) \vee \Big(\overline{f_1(x) \vee f_2(x) \vee f_3(x)}\Big)\bigg)$$

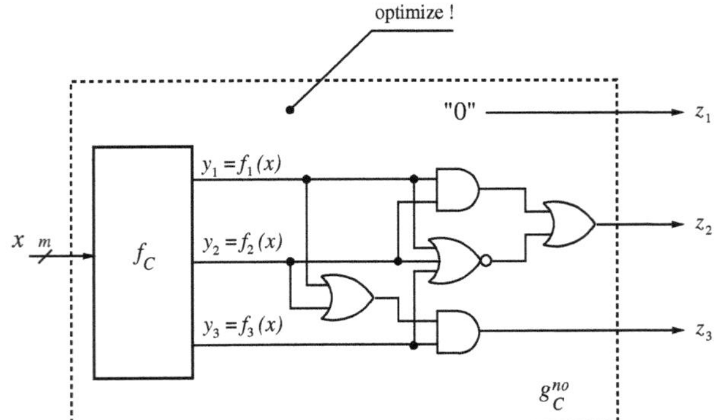

Fig. 3.35 Complementary circuit for a *1-out-of-3* code not yet optimized

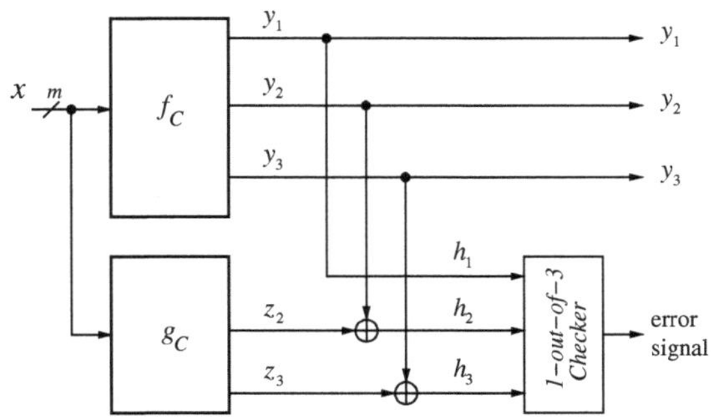

Fig. 3.36 Error detection by a complementary circuit for a *1-out-of-3* code

and

$$h_3(x) = f_3(x) \oplus z_3(x) =$$
$$= f_3(x) \oplus \Big(\big(f_1(x) \vee f_2(x)\big) \wedge f_3(x)\Big) = f_3(x) \wedge \big(\overline{f_1(x) \vee f_2(x)}\big).$$

The output $z_1(x) = 0$ does not have to be implemented.

Now we show that $\big(h_1(x), h_2(x), h_3(x)\big)$ is always a *1-out-of-3* code word. We consider the following cases:

1. $f_1(x) = 1$.
   Then we have
$$h_1(x) = f_1(x) = 1$$
and
$$h_2(x) = f_2(x) \oplus \big(f_2(x) \vee 0\big) = f_2(x) \oplus f_2(x) = 0,$$
$$h_3(x) = f_3(x) \wedge 0 = 0.$$

2. $f_1(x) = 0$.
   Then we conclude
$$h_1(x) = f_1(x) = 0$$
and
$$h_2(x) = f_2(x) \oplus \big(\overline{f_2(x) \vee f_3(x)}\big),$$
$$h_3(x) = f_3(x) \wedge \overline{f}_2(x).$$

There are two possible cases, $f_2(x) = 1$ or $f_2(x) = 0$. $f_2(x) = 1$ implies $h_2(x) = 1$ and $h_3(x) = 0$. For $f_2(x) = 0$ we have $h_2(x) = \overline{f}_3(x)$ and $h_3(x) = f_3(x)$, and either $h_2(x)$ of $h_3(x)$ but not both are equal to 1.

For a circuit with $n, n > 3$ outputs $y_1, \ldots, y_n$ we divide the n outputs into p disjoint groups of 3 outputs $G_j = \{y_{j1}, y_{j2}, y_{j3}\}$, $j = 1, \ldots, p$ and l remaining outputs y_{k1}, y_{kl} with $l \leq 2$.

For every group G_j of three outputs the first output y_{j1} is not complemented and the complementary circuit g_{Cj} determines two complementary outputs g_{j2} and g_{j3} to complement the outputs y_{j2}, y_{j3}.

The remaining l outputs y_{k1}, y_{kl} are duplicated with inverse values.

We expect that the necessary area for the implementation of the complementary output g_{j2} is correlated to the number of ones of the expression

$$f_{j1}(x) \wedge f_{j2}(x).$$

For $j = 1, \ldots, p$ the groups of three outputs are determined as follows according to [63]:

1. As the first output y_{j1} of the group G_j we choose from the outputs of f_C not yet selected an output that is implemented with a maximum number of literals. To this output no complementary output will be added modulo 2.
2. As the second output y_{j2} of G_j we choose an output from the outputs of f_C not yet selected for which under 10000 pseudo random inputs the number of ones of the expression $f_{j1}^{k1}(x) \wedge f_{j2}^{k2}(x)$ with $k1$, $k2 \in \{0, 1\}$ is minimal. Thereby we denote $f_{ji}^0 = f_{ji}$, and $f_{ji}^1 = \overline{f}_{ji}$. For $k1 = 1$ an inverter will be added at the output y_{j1} and for $k2 = 1$ at the output y_{j2}. For $k1 = 0$ ($k2 = 0$) the output y_{j1} (y_{j2}) will not be inverted.
   (We expect that the on-set of $f_{j1}^{k1}(x) \wedge f_{j2}^{k2}(x)$ is small and that this function can be implemented with a small area overhead.)
3. Similarly, we determine the third output of G_j. Again, for 10000 pseudo random inputs we select the output y_{j3} from the outputs not yet selected for which the number of ones in the expression $\left(\overline{f_{j1}^{k1}(x) \vee f_{j2}^{k2}(x)}\right) \wedge f_{j3}^{k3}(x)$ is minimal. For $k3 = 1$ the output y_{j3} will be inverted.
   (We again expect that the on-set of $\left(\overline{f_{j1}^{k1}(x) \vee f_{j2}^{k2}(x)}\right) \wedge f_{j3}^{k3}(x)$ and therefore also the necessary area for the implementation of this function will be small.)

For *1-out-of-4* codes and for *Berger Codes* similar results are described in [61, 62].

3.6.3 Conditions for the Existence of Totally Self-Checking Error Detection Circuits by Complementary Circuits

This section, which is based on [65], describes the conditions necessary and sufficient for the existence of totally self-checking error detection circuits. The error detection circuits are designed by use of complementary circuits for a *1-out-of-n* code.

First we consider a combinational circuit f_C with three inputs $(x_1, x_2, x_3) = x$ and four outputs $(y_1, y_2, y_3, y_4) = y$. The truth table for the Boolean functions

$$y_1 = f_1(x),$$
$$y_2 = f_2(x),$$
$$y_3 = f_3(x),$$
$$y_4 = f_4(x)$$

implemented at the outputs of f_C is presented in Table 3.6.

A totally self-checking checker cannot be designed in the traditional way for the circuit f_C for either a systematic code or for a non-systematic code.

Since for every input one or two outputs are equal to 1 the circuit f_C can be checked by means of a *2-out-of-5* code. As shown in Table 3.6 a single additional output z with

$$z = \overline{x}_1 \vee x_2 x_3 \vee \overline{x}_2 \overline{x}_3$$

has to be added. The possible output vectors are now (1,0,0,0,1), (0,0,0,1,1), (0,1,0,0,1) and (0,0,1,1,0).

These four output vectors are the input vectors of the *2-out-of-5 checker*. The length of a complete test for single stuck-at faults for an arbitrary *m-out-of-n checker* has to be at least t, $t \leq n$ with $n = 5$ [66].

Since only four different input vectors are available at the inputs of the *2-out-of-5 checker* this checker is not completely tested during normal operation.

For a Berger code the number of different input vectors for the *Berger code checker* is also too small to guarantee the self-testing property of the checker [67].

If the circuit f_C is checked by use of a linear systematic code, the check bits of that code are determined by some XOR-trees from the outputs of f_C by the generator circuit. At least two binary outputs of f_C have to be directly XORed by an XOR-gate XOR'. To test XOR' all the four possible input combinations 00, 01, 10 and 11 must be applied at its inputs. From Table 3.6 it can be seen by inspection that for every pair of outputs of f_C only three different input combinations occur. There is no pair

Table 3.6 Example for which no traditional totally self-checking checker exists

	inputs			*outputs*				
	x_1	x_2	x_3	f_1	f_2	f_3	f_4	z
1	0	0	0	1	0	0	0	1
2	0	0	1	0	0	0	1	1
3	0	1	0	0	0	0	1	1
4	0	1	1	0	1	0	0	1
5	1	0	0	0	1	0	0	1
6	1	0	1	0	0	1	1	0
7	1	1	0	0	0	1	1	0
8	1	1	1	1	0	0	0	1

of outputs of f_C which outputs all the four input combinations 00, 01, 10 and 11 needed to test the gate XOR'.

Contrary to the result that no totally self-checking checker can be designed for f_C by the well-known traditional methods, a totally self-checking checker can be designed by a complementary circuit for a *1-out-of-4* code.

The complementary circuit g_C with four outputs z_1, z_2, z_3, z_4 implements the Boolean functions g_1, g_2, g_3, g_4 such that

$$h_1(x) = f_1(x) \oplus g_1(x),$$

$$h_2(x) = f_2(x) \oplus g_2(x),$$

$$h_3(x) = f_3(x) \oplus g_3(x)$$

and

$$h_4(x) = f_4(x) \oplus g_4(x)$$

are elements of a *1-out-of-4* code.

Table 3.7 is the truth table for $f_1, f_2, f_3, f_4, g_1, g_2, g_3, g_4, h_1, h_2, h_3, h_4$.

All the XOR-gates XORing $f_i \oplus g_i$ for $i = 1, \ldots, 4$ are completely tested since all four possible input values 00, 01, 10, 11 occur at the output pairs f_i, g_i and also all the possible *1-out-of-4* code words 1000, 0100, 0010, 0001 are generated at the outputs of the complemented circuit as inputs of the *1-out-of-4* checker.

For this example it was shown that a totally self-checking checker can be designed by a complementary circuit although for the traditional methods a totally self-checking checker does not exist.

Now we describe the necessary and sufficient conditions for the existence of a totally self-checking checker for a combinational circuits f_C by use of a complementary circuit and a *1-out-of-n* code as shown in Fig. 3.37.

The circuit f_C has n outputs $y_1, \ldots, y_r, y_{r+1}, \ldots, y_n$. The first outputs $y_1, \ldots, y_r$ are complemented by the r outputs $z_1, \ldots, z_r$ of the complementary circuit g_C. The remaining $n - r$ outputs $y_{r+1}, \ldots, y_n$ of f_C are not complemented.

For $i = 1, \ldots, r$ the output y_i of f_C and z_i of g_C is XORed into h_i by the XOR-gate XOR_i.

Table 3.7 Example of Table 3.6 with a totally self-checking checker by a complementary circuit

	x_1	x_2	x_3	f_1	g_1	f_2	g_2	f_3	g_3	f_4	g_4	h_1	h_2	h_3	h_4
1	0	0	0	1	0	0	0	0	0	0	0	1	0	0	0
2	0	0	1	0	0	0	0	0	0	1	0	0	0	0	1
3	0	1	0	0	0	0	0	0	1	1	1	0	0	1	0
4	0	1	1	0	0	1	0	0	0	0	0	0	1	0	0
5	1	0	0	0	0	1	1	0	0	0	1	0	0	0	1
6	1	0	1	0	0	0	0	1	0	1	1	0	0	1	0
7	1	1	0	0	1	0	0	1	1	1	1	1	0	0	0
8	1	1	1	1	1	0	1	0	0	0	0	0	1	0	0

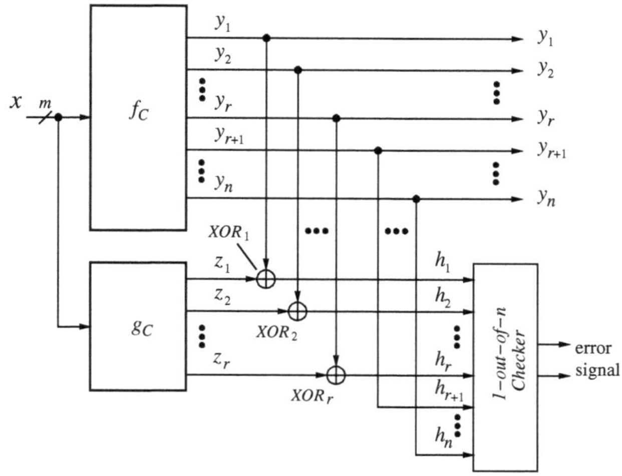

Fig. 3.37 Complementary circuit for a *1-out-of-n* code

The necessary and sufficient conditions for the existence of a totally self-checking checker are expressed by the following

Theorem 3.1. [65] *Let f_C be a combinational circuit implementing at its n outputs $y_1,\ldots,y_n$ the Boolean functions $y_1 = f_1(x),\ldots,y_n = f_n(x)$ where all these Boolean functions are not constant for $x \in \chi$ and where χ is the set of inputs actually applied to f_C during normal operation.*

It is then possible to design a totally self-checking error detection circuit by use of a complementary circuit g_C for a 1-out-of-n *code where the complementary circuit g_C implements at its r, $2 \le r \le n$, outputs $z_1,\ldots,z_r$ Boolean functions $z_1 = g_1(x),\ldots,z_r = g_r(x)$ such that the $n-r$ Boolean functions $y_{r+1} = f_{r+1}(x),\ldots,y_n = f_n(x)$ are not complemented and that the outputs $y_1 = f_1(x),\ldots,y_r = f_r(x)$ are complemented by the r outputs $z_1 = g_1(x),\ldots,z_r = g_r(x)$ if and only if the following conditions are satisfied:*

1. *For $n > r+1$ the non-complemented functions f_i, f_j with $i,j \in \{r+1,\ldots,n\}$, $i \neq j$ are pairwise orthogonal,*

$$\sum_{x\in\chi} f_i(x)f_j(x) = 0 \quad for \quad i,j \in \{r+1,\ldots,n\} \quad and \quad i \neq j,$$

2. *Let $\widetilde{\chi}$ be the subset of inputs for which one of the non-complemented functions $y_{r+1} = f_{r+1}(x),\ldots,y_n = f_n(x)$ is equal to 1,*

$$\widetilde{\chi} = \left\{x \in \chi, \quad \exists i \in \{r+1,\ldots,n\}, \quad with \quad f_i(x) = 1\right\}.$$

Then for every complemented output $y_i = f_i(x), \quad i \in \{1,\ldots,r\}$ of f_C there exists a subset of inputs

$$\kappa_i = \{x_i^1, x_i^0\} \subset \chi \setminus \widetilde{\chi} \quad with \quad f_i(x_i^1) = 1, f_i(x_i^0) = 0$$

and

$$\kappa_i \cap \kappa_j = \emptyset \quad for \quad i, j \in \{1, \ldots, r\} \quad and \quad i \neq j.$$

3. *For every* $i = 1, \ldots, r$ *we have*

$$|On(f_i)| \leq 2 \quad and \quad |Off(f_i)| \leq 2.$$

Proof:

We prove here that only the conditions of Theorem 3.1 are necessary. That these conditions are also sufficient is shown in [65, 68].

To detect all single stuck-at faults of the *1-out-of-n checker* all the *n 1-out-of-n* code words have to be applied to the checker and therefore to be generated at the outputs $h_1, \ldots, h_n$ of the complemented circuit.

To test all the XOR-gates $XOR_1, \ldots, XOR_r$ all the four inputs 00, 01, 10, 11 have to be applied to these XOR-gates.

It is obvious that the non-complemented outputs $y_{r+1} = f_{r+1}(x), \ldots, y_n = f_n(x)$ have to be orthogonal. Otherwise a non-code word would be generated for some input.

Now we distinguish whether the inputs are from $\widetilde{\chi}$ or from $\chi \setminus \widetilde{\chi}$.

1. Let $x \in \widetilde{\chi}$.
   Then exactly one of the outputs

$$y_{r+1} = f_{r+1}(x),$$
$$\vdots$$
$$y_n = f_n(x)$$

is equal to 1 and we have

$$y_1 = h_1(x) = 0,$$
$$\vdots$$
$$y_r = h_r(x) = 0.$$

Otherwise no *1-out-of-n* code word will be generated. For $i = 1, \ldots, r$ we have

$$h_i(x) = f_i(x) \oplus g_i(x) = 0$$

or

$$f_i(x) = g_i(x)$$

and the XOR-gate XOR_i will not be tested by 01 and 10.

2. Let now $x \in \chi \setminus \widetilde{\chi}$.
   Then the non-complemented outputs

$$y_{r+1} = f_{r+1}(x),$$
$$\vdots$$
$$y_n = f_n(x)$$

are all equal to 0.

One of the complemented outputs, say $h_i(x) = f_i(x) \oplus g_i(x)$, $i \in \{1,\dots,r\}$ has to be 1, and all the other complemented outputs $h_j(x) = f_j(x) \oplus g_j(x)$, $j \in \{1,\dots,r\}$ with $i \neq j$ have to be 0.
This is only possible for

$$f_i(x) = 1 \quad \text{and} \quad g_i(x) = 0$$

or for

$$f_i(x) = 0 \quad \text{and} \quad g_i(x) = 1$$

and for $f_j(x) = g_j(x), \quad i \neq j, \quad i,j \in \{1,\dots,r\}$.

To test the XOR-gate XOR_i by 10 and 01 we conclude that there must exist two different inputs x_i^1 and x_i^0, with $x_i^1, x_i^0 \in \chi \setminus \widetilde{\chi}$ and

$$f_i(x_i^1) = \overline{g}_i(x_i^1) = 1$$

and

$$f_i(x_i^0) = \overline{g}_i(x_i^0) = 0.$$

For $i \neq j$ we have

$$f_j(x_i^1) = g_j(x_i^1),$$
$$f_j(x_i^0) = g_j(x_i^0).$$

We set $\kappa_i = \{x_i^1, x_i^0\}$.

Since all the r *1-out-of-n* code words

$$\underbrace{1\,0\dots0}_{r}\underbrace{0\dots0}_{n-r},$$
$$\vdots$$
$$\underbrace{0\dots0\,1}_{r}\underbrace{0\dots0}_{n-r}$$

have to be generated for some $x \in \chi \setminus \widetilde{\chi}$, all the XOR-elements $XOR_1,\dots,XOR_r$ have to be tested and the r sets

$$\kappa_1 = \{x_1^1, x_1^0\},$$
$$\vdots$$
$$\kappa_r = \{x_r^1, x_r^0\}$$

have to exist.

Since for $x \in \kappa_i$ we have $f_i(x) = \overline{g}_i(x)$ and $f_j(x) = g_j(x)$ for $i \neq j$ the sets $\kappa_1, \ldots, \kappa_r$ are mutually disjoint.

To guarantee that the XOR-elements $XOR_1, \ldots, XOR_r$ are also tested by 00 and 11 for $i \in \{1, \ldots, r\}$ there have to exist besides the inputs x_i^1 and x_i^0 with $f_i(x_i^1) = \overline{g}_i(x_i^1) = 1$ and $f_i(x_i^0) = \overline{g}_i(x_i^0) = 0$ two additional inputs $\tilde{x}_i^1, \tilde{x}_i^0$ with

$$f_i(\tilde{x}_i^1) = g_i(\tilde{x}_i^1) = 1$$

and

$$f_i(\tilde{x}_i^0) = g_i(\tilde{x}_i^0) = 0.$$

This implies $|On(f_i)| \leq 2$ and $|Off(f_i)| \leq 2$, which finishes the proof that the conditions are necessary.

Simple sufficient conditions for the existence of a totally self-checking error detection circuit by use of a *1-out-of-n* code are given in Theorem 3.2.

Theorem 3.2. [65] *Let f_C be a combinational circuit with the set χ of actually applied inputs implementing at its n outputs $y_1, \ldots, y_n$ the Boolean functions $y_1 = f_1(x), \ldots, y_n = f_n(x)$ and let*

$$X_1^0 = \{x, x \in \chi \wedge f_1(x) = 0\},$$
$$X_1^1 = \{x, x \in \chi \wedge f_1(x) = 1\},$$
$$\vdots$$
$$X_n^0 = \{x, x \in \chi \wedge f_n(x) = 0\},$$
$$X_n^1 = \{x, x \in \chi \wedge f_n(x) = 1\}.$$

If the subsets $X_1^0, X_1^1, \ldots, X_n^0, X_n^1$ are ordered with respect to their number of elements in ascending order as $X_1', X_2', \ldots, X_{2n}'$ and if $|X_1'| \geq 2$ and $|X_k'| \geq k$ for $k = 2, \ldots, 2n$ then there exists a totally self-checking error detection circuit by use of a complementary circuit for a 1-out-of-n *code.*

Proof:
We assume that all the outputs of f_C are complemented and we show that the conditions 2 and 3 of Theorem 3.1 are valid.

We determine for $j = 1, \ldots, n$ the set κ_j.

Let $X_1' = X_{i1}^{c_{i1}}$ where $c_{i1} \in \{0, 1\}$. Then we choose an arbitrary element $x_{i1} \in X_{i1}^{c_{i1}}$ as an element of κ_{i1} with $f_{i1}(x_{i1}) = c_{i1}$.

Now we remove the set $X_1' = X_{i1}^{c_{i1}}$ from the sets $X_1', X_2', \ldots, X_{2n}'$, we delete the input x_{i1} from all the remaining sets $X_2', \ldots, X_{2n}'$, we order the remaining sets with the already deleted element x_{i1} again with respect to their numbers of elements in ascending order and we obtain $X_2', \ldots, X_{2n}'$.

Let now $X_2' = X_{i2}^{c_{i2}}$ where $c_{i2} \in \{0,1\}$. Then we choose an arbitrary element $x_{i2} \in X_{i2}^{c_{i2}}$ as an element of κ_{i2} with $f_{i2}(x_{i2}) = c_{i2}$.

Now we remove the set $X_2' = X_{i2}^{c_{i2}}$ from the sets $X_2', \ldots, X_{2n}'$, we delete the input x_{i2} from all the remaining sets $X_3', \ldots, X_{2n}'$, we order the remaining sets with the already deleted element x_{i2} again with respect to their numbers of elements in ascending order and we obtain $X_3', \ldots, X_{2n}'$.

We continue until all the sets X_j' are removed and all the sets $\kappa_1, \kappa_2, \ldots, \kappa_{2n}$ are determined. We have $\kappa_i \cap \kappa_j = \emptyset$ for $i \neq j$ and $|\kappa_i| = 2$ for $i = 1, \ldots, 2n$ by construction and the condition 2 of Theorem 3.1 is fulfilled.

Since we assumed $|X_1'| \geq 2$ and $|X_k'| \geq k$ for $k = 2, \ldots, 2n$ also condition 3 is valid, which concludes the proof.

For practical applications for circuits with a large number of inputs the exact sets X_1^0, $X_1^1, \ldots$, X_n^0, X_n^1 may be replaced by subsets of theses sets determined for $N \approx 10000$ pseudo random inputs.

In this section the method of error detection by complementary circuits was described.

The definition of a complementary circuit for a given functional circuit and a considered code was given and it was explained how a complementary circuit can be determined such that the componentwise *XOR*-sum of the outputs of the functional and of the complementary circuit is a code word of a chosen code as long as no error occurs.

It was demonstrated that almost arbitrary codes such as *m-out-of-n* codes, Berger codes, systematic and non-systematic linear codes can be used for the design of complementary circuits. Therefore the error detection circuits can be easily designed in accordance with the codes which are already used in the system. No code adaptation is necessary.

It was demonstrated that a complementary circuit for a completely defined functional circuit is only "partially specified" since for every input the *XOR*-sum of the outputs of the functional circuit and of the complementary circuit must not be a specific, but an arbitrary code word. It was shown that for a functional circuit with n inputs and a code with N code words there are N^{2^n} different complementary circuits from which the "optimum" can be determined since for all these complementary circuits the *XOR*-sum of the outputs of functional and of the complementary circuit is a code word of the considered code. It remains a challenging synthesis problem to utilize this plurality of possible specification of the complementary circuit for a systematic circuit optimization. This optimization problem cannot be handled until now by the available synthesis tools. Therefore heuristic solutions are needed, which were described for the special case of a *1-out-of-3* code.

It was explained that the problem of circuit optimization is different for a complementary circuit compared to the traditional method of error detection by use of systematic codes. In the latter case the predictor and generator circuits for the considered code are completely specified and can be synthesized by a CAD tool from the netlist of the functional circuit.

For the example of a *1-out-of-3* code, heuristic solutions were described in detail. It was explained how the outputs of the functional circuit can be divided into groups of 3 outputs and how the corresponding outputs of the complementary circuit can be determined.

It was shown that there are functional circuits for which no totally self-checking checker can be designed using the traditional error detection methods, but for which a totally self-checking checker can be designed by use of a complementary circuit.

For an arbitrarily given functional circuit necessary and sufficient conditions for the existence of a self-checking checker are proven for a *1-out-of-n* code and also simple sufficient conditions were derived.

It was shown that these conditions are fulfilled for almost all circuits.

It was demonstrated that the method of error detection by complementary circuits is a new method of error detection with many interesting properties. This method needs further investigation.

3.7 General Method for the Design of Error Detection Circuits

In this section we will show how the design of an optimum error detection circuit can be reduced to a classical design problem for a partially defined combinational circuit. Only combinational circuits as functional circuits will be considered in this section. For sequential circuits a similar method can be found in [69, 70].

It will be explained how a partially defined Boolean function, which is called the error detection function, can be determined from the the functional circuit and from the considered fault model. An optimum implementation of this partially defined error detection function is the optimum error detection circuit. This error detection circuit detects all the errors caused by the faults of the considered fault model.

It will be explained how the known methods of error detection - duplication and comparison, error detection by codes and error detection by complementary circuits - are special cases of this general approach.

In summary it will be shown that in principle the design of optimum error detection circuits can be reduced to a standard CAD design problem. But the special solutions described in the previous chapters of this book remain of interest since until now the available synthesis tools do not generate better solutions for circuits of realistic sizes.

3.7.1 Description of the Method

We now describe how an optimum error detection circuit for a given combinational circuit with a given error model can be designed.

We consider an arbitrarily given combinational circuit f_C with m binary inputs $x = (x_1, \ldots, x_m)$ and n binary outputs $y = (y_1, \ldots, y_n)$.

The combinational circuit f_C implements at its outputs the n Boolean functions

$$y_1(x) = f_1(x),$$
$$\vdots$$
$$y_n(x) = f_n(x)$$

and we denote

$$y(x) = y_1(x), \ldots, y_n(x)$$

and

$$f(x) = f_1(x), \ldots, f_n(x).$$

The set χ, with $\chi \subseteq X = \{0,1\}^m$ is the set of inputs for which the behavior of f_C is of interest. For $\chi \subset X$ not all the possible inputs are really applied during normal operation.

$\Phi = \{\varphi_1, \ldots, \varphi_L\}$ is the set of considered faults.

In the presence of a fault $\varphi \in \Phi$ the outputs of f_C under input $x \in \chi$ are denoted by

$$\begin{aligned} y(\varphi,x) &= y_1(\varphi,x), \ldots, y_n(\varphi,x) = \\ &= f_1(\varphi,x), \ldots, f_n(\varphi,x) = f(\varphi,x). \end{aligned}$$

Now we define the *error detection function F(x,y)* by

$$F(x,y) = \begin{cases} 0 & \text{for} \quad x \in \chi \quad \text{and} \quad y = f(x), \\ 1 & \text{for} \quad x \in \chi \quad \exists \varphi \in \Phi, \quad \text{with} \quad y = f(\varphi,x) \neq f(x), \\ \sim \text{(don't care)} & \text{otherwise.} \end{cases}$$

The error detection function $F(x,y)$ is equal to 0 if x is from the expected input set χ and if the output of f_C is correct. $F(x,y)$ is equal to 1 if

1. x is from the expected input set $\chi \subseteq X$
   and
2. there exists a fault φ from the considered set of faults Φ such that $y = f(\varphi,x)$, and
3. if $y = f(\varphi,x)$ is different from the expected correct output $f(x)$.

In all other cases, i.e. for all inputs that are not from the expected input set χ, and for all faults that are not from the fault model Φ of the considered faults, the error detection function F(x,y) is "don't care".

Here it is assumed that, in the presence of a fault, the faulty circuit remains combinational.

An "optimum" implementation of the error detection function $F(x,y)$ by an available synthesis tool is an "optimum" error detection circuit for the given circuit f_C and the given fault model Φ.

Thus, for an arbitrarily given combinational circuit f_C and a given fault model $\Phi = \{\varphi_1, \ldots, \varphi_L\}$ the design of an "optimum" error detection circuit F_C can in principle be reduced to an optimum implementation of a partially defined error detection function $F(x,y)$.

The block diagram of the functional circuit f_C monitored by the error detection circuit F_C is shown in Fig. 3.38.

The standard methods of error detection, duplication and comparison, error detection with predictor and generator circuits, error detection by complementary circuits are all special cases of this general approach.

This will be explained by the following Figures.

Figure 3.39 shows error detection by duplication and comparison. The error detection circuit F_C here consists of the duplicated circuit f_{C_d} and the *comparator*.

Figure 3.40 presents error detection by systematic block codes. The predictor $Pr(x)$, the generator G and the *comparator* form the error detection circuit F_C.

In the case of error detection by complementary circuits in Fig. 3.41 the error detection circuit F_C consists of the complementary circuit g_C, the XOR-gates XORing for $i = 1, \ldots, n$ the outputs y_i of f_C and z_i of g_C and the *code checker*.

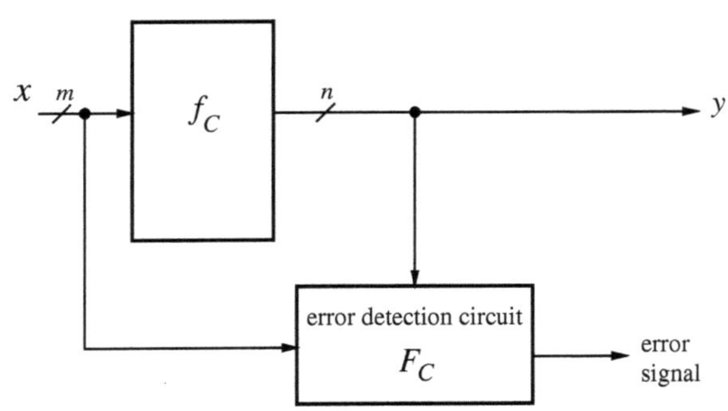

Fig. 3.38 General method of error detection for a combinational circuit f_C

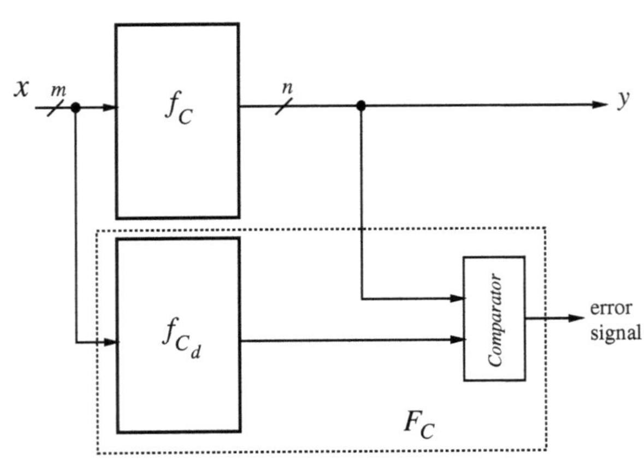

Fig. 3.39 Special case duplication and comparison

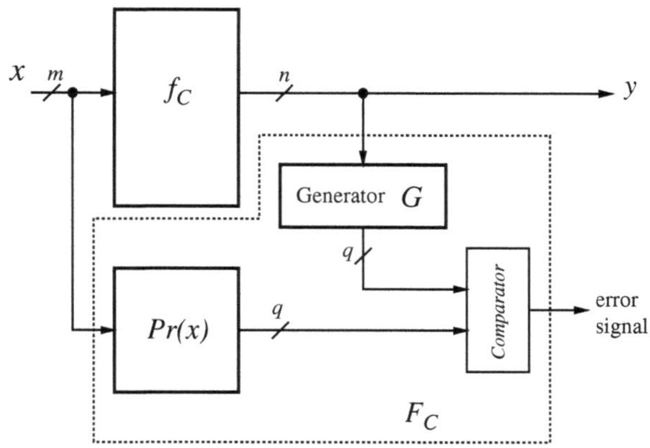

Fig. 3.40 Special case error detection by systematic block codes

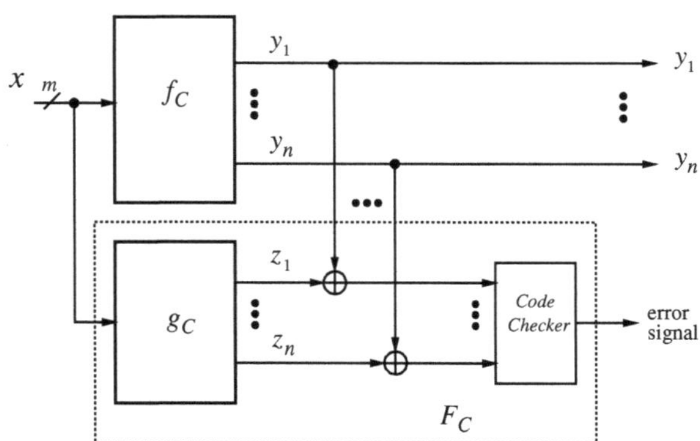

Fig. 3.41 Special case error detection by complementary circuits

Although in principle the design of an optimum error detection circuit can be reduced to a standard problem of CAD, the optimum design of a partially specified Boolean function, special human designs are necessary. Although synthesis tools are improving, the available CAD tools are not sufficiently adequate to generate satisfactory solutions. This is a common experience and is emphasized in [71].

In this section it was shown how an optimum error detection circuit can be designed as an optimum implementation of a partially defined error detection function. It was explained how this error detection function is determined by the functional circuit and by the considered fault model chosen by the designer.

It was demonstrated that the known methods of error detection, such as duplication and comparison, error detection by systematic codes and error detection by use of complementary circuits, are special cases of this general approach.

3.8 Self-Dual Error Detection

In this section the method of error detection by means of *self-dual Boolean functions* will be described. Self-dual Boolean functions are implemented as *self-dual combinational circuits*.

It is shown how a Boolean function, if it is not self-dual, can be transformed into a self-dual Boolean function and can be implemented as a "self-dual combinational circuit".

For error detection *alternating inputs*, i.e. the functional inputs and their corresponding componentwise inverted inputs are applied to the self-dual circuit. For alternating inputs the circuit outputs of these self-dual circuits are alternating. If, due to a fault in the self-dual circuit or at the input lines the outputs are not alternating, a fault is detected.

Since in addition to the original inputs the corresponding alternating inputs (i.e. the componentwise inverted inputs) are always subsequently applied to the circuit inputs time redundancy is 100%, and the described method can only be applied if time is not critical. This is often the case, for instance, in mechanical control systems.

Two different methods for the transformation of an arbitrarily given combinational circuit in a self-dual circuit are described. Here it is assumed that the circuits are given as a netlist of gates.

In the first method, which was already introduced in [72], an additional input variable is used.

In the second method according to [73] for the given combinational circuit a complementary circuit is determined such that the componentwise *XOR*-sums of the corresponding outputs of the original circuit and of the complementary circuit are self-dual.

The componentwise *XOR*-sum of the outputs of the original circuit and of the complementary circuit can be an arbitrary self-dual Boolean function. Therefore, for a given functional circuit there are many different possibilities to determine a complementary circuit. To utilize all these possibilities for circuit optimization remains a challenging synthesis problem which is not yet considered in the available synthesis tools. Some heuristic methods alone are described in this chapter.

Error detection by self-dual parity and self-dual duplication are considered in more detail.

In ordinary parity prediction the parity of the outputs, which is determined by an *XOR*-tree from the circuit outputs, is compared with the predicted parity, which is determined from the circuit inputs.

In self-dual parity checking the predicted parity of ordinary parity checking is replaced by a complementary function of the parity function such that the *XOR*-sum of the complementary function and the output of the *XOR*-tree of the circuit outputs is an arbitrary self-dual Boolean function. An optimum complementary function can be selected from the large variety of possible self-dual complements.

Similarly, in self-dual duplication the duplicated circuit is replaced by a complementary circuit (with the same number of outputs as the original circuit), such that the componentwise *XOR*-sums of the outputs of the original circuit and of the complementary circuit are all self-dual.

Separate and joint implementations of the original circuit and of the complementary circuit are also considered.

At the end of this section self-dual fault-secure circuits are introduced. It is shown how a self-dual circuit can be transformed into a self-dual fault-secure circuit by a simple circuit transformation.

Self-dual circuits can be operated in

1. a fast mode without error detection in which only the functional inputs and not the alternating inputs are applied,
2. a slow error detection mode in which the functional inputs and also the alternating inputs are always applied
3. and in a self-test mode in which alternating inputs are applied only during the test.

3.8.1 Self-Dual Boolean Functions

First we recapitulate the definition of a self-dual Boolean function.

Definition 3.10. Let $f(x)$, $x = x_1,\ldots,x_m$ be a Boolean function of length m. Then $\overline{f}(\overline{x})$ is called the *dual function* $f_d(x)$ of $f(x)$,

$$f_d(x) = \overline{f}(\overline{x}).$$

Definition 3.11. A Boolean function $f(x)$ is *self-dual* if its dual function $f_d(x)$ is equal to the original function $f(x)$, i.e. if

$$f(x) = f_d(x) = \overline{f}(\overline{x}),$$

or if

$$f(\overline{x}) = \overline{f}(x).$$

Figure 3.42 shows a combinational circuit $f_{C_{sd}}$ implementing a self-dual Boolean function $f(x)$. Alternating inputs x and $\overline{x}$ are applied. The corresponding outputs of $f_{C_{sd}}$ are $y = f(x)$ and $\overline{y} = f(\overline{x})$ and the outputs of $f_{C_{sd}}$ are alternating for alternating inputs.

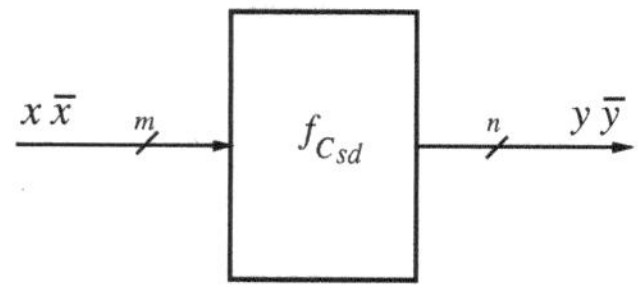

Fig. 3.42 Behavior of self-dual circuit

Definition 3.12. A circuit f_C implementing at its n outputs, $n \geq 1$, the Boolean functions

$$
\begin{aligned}
y_1(x) &= f_1(x), \\
&\vdots \\
y_n(x) &= f_n(x)
\end{aligned}
$$

is called *self-dual* if all the Boolean functions $f_1, \ldots, f_n$ are self-dual.

3.8.2 Transformation of a Given Circuit into a Self-Dual Circuit

Now we describe two different methods how a combinational circuit can be transformed into a self-dual circuit.

3.8.2.1 Self-Dual Circuits by Use of an Additional Input

The first method is based on the fact that an arbitrarily given Boolean function $f(x)$ can be transformed into a self-dual function $F_{sd}(a,x)$ by use of an additional binary variable a according to [72]. The self-dual function $F_{sd}(a,x)$ is determined as

$$F_{sd}(a,x) = \overline{a} f(x) \vee a \overline{f}(\overline{x}) \tag{3.30}$$

with $a \in \{0,1\}$, and we have

$$
\begin{aligned}
f(x) &= F_{sd}(0,x) \\
&\text{and} \\
\overline{f}(\overline{x}) &= F_{sd}(1,x).
\end{aligned}
$$

It is easy to show by direct calculation that $F_{sd}(a,x)$ is a self-dual Boolean function.

Usually a circuit f_C implementing the Boolean function $f(x)$ is given as a netlist of gates. If f_C is implemented by *AND*-gates, *OR*-gates and *INVERTERS* and if all the *AND*-gates are replaced by *OR*-gates and all the *OR*-gates by *AND*-gates the transformed circuit f_{Cd} will implement the dual function $f_d(x) = \overline{f}(\overline{x})$ [74].

Figure. 3.43 illustrates the design of a self-dual circuit according to this method. The circuitry of Fig. 3.43 consisting of the original functional circuit f_C, the dual circuit f_{C_d} derived from f_C, the *AND* and *OR*-gates with the additional input line a have to be optimized by an available synthesis tool.

Functionally considered, the self-dual circuit in Fig. 3.43 is uniquely specified. The binary variable a selects as a control signal whether the outputs of the

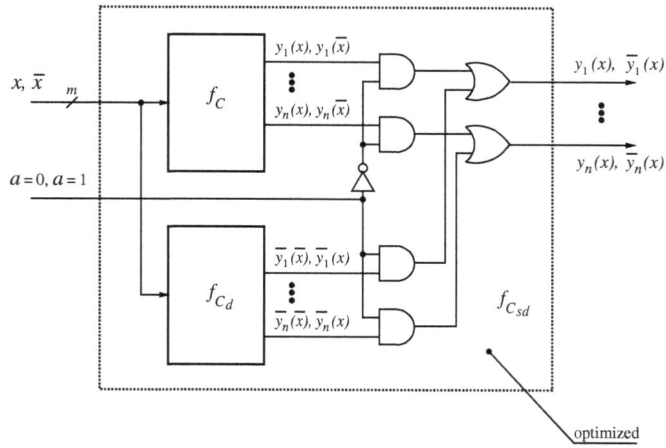

Fig. 3.43 Transformation of a combinational circuit into a self-dual circuit

original circuit f_C or the dual circuit f_{C_d} are connected to the circuit outputs. First the input $(x, a = 0)$ is applied, and the outputs $y_1(x), \ldots, y_n(x)$ of f_C are connected to the circuit outputs. Next, for the alternating input $(\overline{x}, a = 1)$ the outputs $\overline{y}_1(\overline{x}), \ldots, \overline{y}_n(\overline{x}) = \overline{y}_1(\overline{\overline{x}}), \ldots, \overline{y}_n(\overline{\overline{x}})$ of f_{C_d} are connected to the circuit outputs, and for alternating inputs the outputs are also alternating.

3.8.2.2 Self-dual Circuits by Use of Complementary Circuits

A Boolean function $f(x)$ can also be transformed into a self-dual function by use of a self-dual complement [75].

Definition 3.13. Let $f(x)$, $x = x_1, \ldots, x_m$ be a Boolean function. Then the Boolean function $\delta_f(x)$ is a self-dual complement of $f(x)$ if the Boolean function $h(x)$, with

$$h(x) = f(x) \oplus \delta_f(x) \tag{3.31}$$

is self-dual.

It is well known that for $x = x_1, \ldots, x_m$ there exist $2^{2^{m-1}}$ different self-dual functions $f_{sd}(x)$ of length m.

This implies that *for an arbitrarily given Boolean function $f(x)$ of length m, $x = x_1, \ldots, x_m$ there also exist $2^{2^{m-1}}$ different self-dual complements δ_f.*

It is a challenging synthesis problem to determine from these $2^{2^{m-1}}$ different self-dual complements the "optimum" one which results in a minimal area overhead for the implementation of a self-dual circuit by use of a complementary circuit.

It can be proved by direct calculation that for $i = 1,\ldots,m$

$$\delta_i(x) = x_i \wedge \left(f(x) \oplus \overline{f}(\overline{x})\right) \tag{3.32}$$

is a self-dual complement of $f(x)$.

For these complements the number of input values x for which $\delta_i(x)$ is equal to 1 is minimal [75].

Also for $i = 1,\ldots,m$

$$\tilde{\delta}_i(x) = x_i \vee \left(f(x) \oplus f(\overline{x})\right) \tag{3.33}$$

is a self-dual complement of $f(x)$.

The number of input values x for which $\tilde{\delta}_i(x)$ is equal to 1 is maximal.

These self-dual complements $\delta_i(x)$ and $\tilde{\delta}_i(x)$ can be determined for a circuit f_C which is given as a netlist of gates.

Figure 3.44 illustrates the determination of the circuit δ_{i_C} implementing the self-dual complement $\delta_i(x) = x_i \wedge \left(f(x) \oplus \overline{f}(\overline{x})\right)$ from the original circuit f_C.

We suppose for the simplicity of presentation that f_C has a single output only.

From the original circuit f_C (given as a netlist of *AND*, *OR*-gates and inverters) the dual circuit f_{C_d} is derived by changing all *AND*-gates into *OR*-gates and all *OR*-gates into *AND*-gates. The inputs $x = x_1,\ldots,x_m$ are applied in parallel to both f_C and f_{C_d} and the outputs of f_C and f_{C_d} are XORed. The output of that XOR-gate is connected to the first input of an *AND*-gate. The *i*th component x_i of the circuit input is connected to the second input of that *AND*-gate. The circuit of Fig. 3.44 has to be optimized by a synthesis tool.

For the self-dual complements $\tilde{\delta}_i(x) = x_i \vee \left(f(x) \oplus f(\overline{x})\right)$ the circuit $\tilde{\delta}_{i_C}$ can be similarly determined.

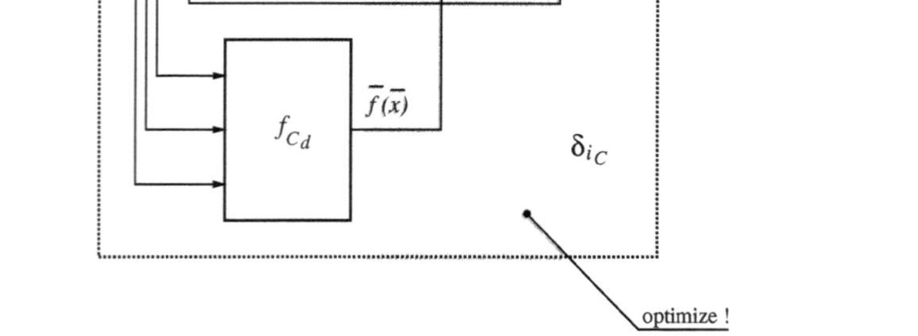

Fig. 3.44 Special self-dual complement δ_{i_C}

3.8.3 Self-Dual Error Detection Circuits

Figure 3.45 shows an error detection circuit by means of a complementary circuit δ_{f_C} for a given combinational circuit f_C implementing a Boolean function $f(x)$. The complementary circuit δ_{f_C} implements a self-dual complement δ_f of f, and the function

$$h(x) = f(x) \oplus \delta_f(x)$$

has to be one of the $2^{2^{m-1}}$ self-dual functions of length m.

A fault is detected under alternating inputs $(x, \overline{x})$ if

$$h_\varphi(x) = h_\varphi(\overline{x})$$

or if the output h_φ is not alternating where h_φ denotes the implemented function h in the presence of a fault φ.

Unlike other (not code-disjoint) methods, faults at the input lines are also detected with this method.

3.8.3.1 Self-Dual Parity

Now we explain how a self-dual complement can be used for error detection by parity prediction [75].

Again we consider a combinational circuit f_C with n outputs $y = (y_1, \ldots, y_n)$ and m inputs $x = (x_1, \ldots, x_m)$.

Instead of checking f_C by comparing the output parity

$$P(y) = y_1 \oplus y_2 \oplus \ldots \oplus y_n$$

of f_C with the parity $P(x)$ predicted from the inputs $x_1, \ldots, x_m$ we design a self-dual complement δ_P of the parity.

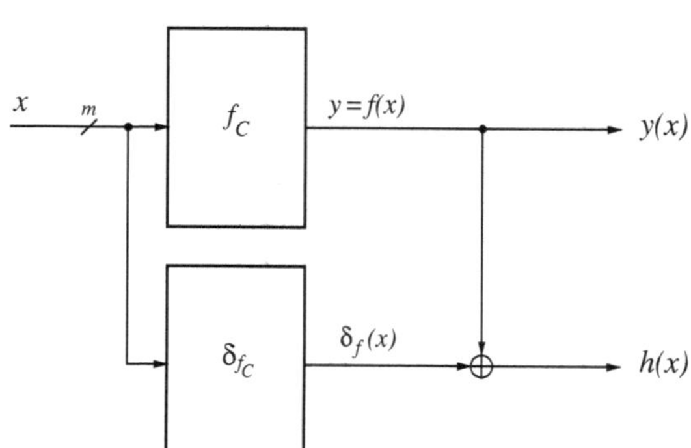

Fig. 3.45 Error detection circuit by use of a self-dual complement

The output parity $P(y)$ and the self-dual complement δ_P are added modulo 2 to form an (arbitrary) self-dual function $h(x)$ of m variables. There are, as already explained, $2^{2^{m-1}}$ possible self-dual complements of $P(y)$ from which the "optimum" has to be selected.

In practice, one of the $2 \times m$ self-dual complements according to equation (3.32) or (3.33) can be selected.

In Fig. 3.46 error detection by self-dual parity is shown for a circuit f_C with $n = 5$ outputs.

As a concrete example we consider a circuit which implements at its outputs the following Boolean functions

$$\begin{aligned}
y_1(x) &= \bar{x}_1 x_2 x_3 \bar{x}_4 \vee (x_1 \vee x_2) x_3 x_4 \vee \bar{x}_1 \bar{x}_3 x_4 \\
y_2(x) &= (\overline{x_1 \vee x_2}) \vee x_1 x_3 \bar{x}_4 \vee x_1 \bar{x}_3 x_4 \\
y_3(x) &= \bar{x}_1 x_2 x_3 \bar{x}_4 \vee (\overline{x_3 \vee x_4}) \vee \bar{x}_1 \bar{x}_2 x_3 \\
y_4(x) &= \bar{x}_1 \bar{x}_2 x_3 \vee \bar{x}_1 \bar{x}_3 x_4 \vee x_1 \bar{x}_3 x_4 \vee x_1 x_3 \bar{x}_4 \\
y_5(x) &= (\overline{x_3 \vee x_4}) \vee (x_3 \vee x_4) \bar{x}_1 x_2.
\end{aligned}$$

Then the predicted parity $P(x) = y_1(x) \oplus \ldots \oplus y_5(x)$ is

$$P(x) = \bar{x}_1 \bar{x}_2 \vee \bar{x}_1 \bar{x}_3 x_4 \vee \bar{x}_1 x_3 \bar{x}_4 \vee x_1 x_3 x_4.$$

As a self-dual complement δ_P we select the function

$$\delta_P(x) = x_2 \wedge \left(P(x) \oplus \overline{P}(\bar{x})\right) = x_2 x_3 x_4.$$

which is much more simple than the predicted parity $P(x)$.

To reduce necessary area overhead the self-dual complement δ_P can be jointly implemented with the functional circuit f_C as shown in Fig. 3.46. Thereby the fault-coverage is not significantly reduced.

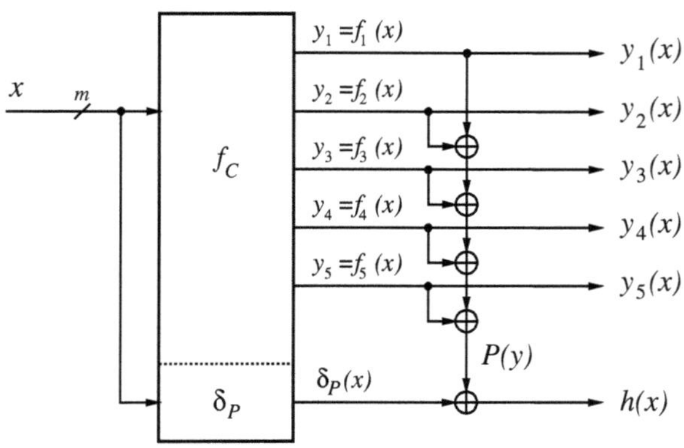

Fig. 3.46 Error detection by self-dual parity

For benchmark circuits the expected area for the implementation of self-dual parity is only 15% of the original circuits and 23% of the optimized circuits compared to 35% and 56% for ordinary parity [75].

3.8.3.2 Self-Dual Duplication

Self-dual duplication is illustrated in Fig. 3.47. For every output $y_1, \ldots, y_n$ of a given functional circuit f_C a self-dual complement $\delta_1, \ldots, \delta_n$ is determined and XORed to form a self-dual function $h_1, \ldots, h_n$.

The circuit δ_{f_C} implementing the self-dual complements $\delta_1, \ldots, \delta_n$ and the functional circuit f_C can be jointly or separately optimized. A joint implementation is impossible for ordinary duplication and comparison.

For benchmark circuits the average area overhead is about 72% of the functional circuit for a separate implementation and about 47% for a joint implementation. Details are described in [76].

3.8.4 Self-Dual Fault-Secure Circuits

In this section we investigate the detectability of stuck-at faults in self-dual circuits.

We introduce the notion of a *self-dual fault-secure* circuit and we show how a given self-dual circuit can be transformed into a "self-dual fault-secure" circuit [77, 75].

We assume that the considered self-dual circuits are given as a netlist of *AND*, *OR*, *NAND*, *NOR*-gates and *INVERTERs*. As faults we consider a set Φ of k single stuck-at faults

$$\Phi = \{\varphi_1, \ldots, \varphi_k\}.$$

A gate G with a stuck-at fault $\varphi \in \Phi$ at an input or an output line will be denoted by G_φ.

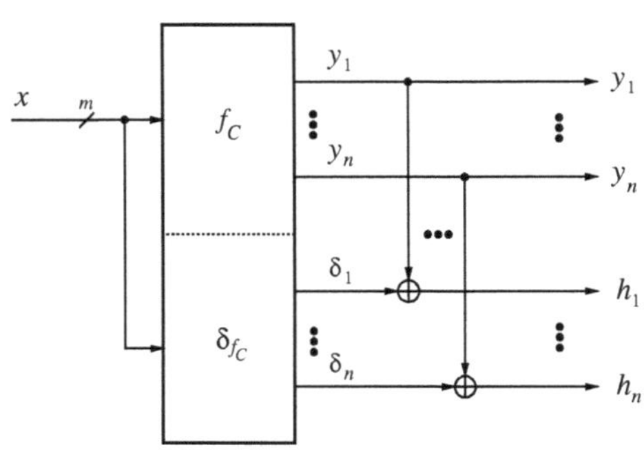

Fig. 3.47 Error detection by self-dual duplication

$G(u)$ and $G_\varphi(u)$ are the Boolean functions implemented by the correct gate G and the faulty gate G_φ, where $u = (u_1, u_2)$ are the input variables of G and G_φ.

If the gate G is an *AND*, *OR*, *NAND*, *NOR*-gate or an *INVERTER* and if a single stuck-at fault φ occurs for G_φ, only one of the following situations is possible:

1. For a subset of inputs for which the output of the correct gate G is 0 the output of the faulty gate G_φ is 1. Then for the subset of inputs for which the output of the correct gate G is 1 the output of the faulty gate G_φ is also 1.
2. For a subset of inputs for which the output of the correct gate G is 1 the output of the faulty gate G_φ is 0. Then for the subset of inputs for which the output of the correct gate G is 1 the output of the faulty gate G_φ is also 0.

More formally this can be expressed by the following

Theorem 3.3. *Let* $U_0 = \{u \in \{0,1\}^2, G(u) = 0\}$ *and* $U_1 = \{u \in \{0,1\}^2, G(u) = 1\}$. *Then we have for a fault* $\varphi \in \Phi$ *either*

1.

$$0 = G(u) \neq G_\varphi(u) \quad for \quad u \in \tilde{U}_0 \subseteq U_0$$

and

$$1 = G(u) = G_\varphi(u) \quad for \quad u \in U_1$$

or

2.

$$1 = G(u) \neq G_\varphi(u) \quad for \quad u \in \tilde{U}_1 \subseteq U_1$$

and

$$0 = G(u) = G_\varphi(u) \quad for \quad u \in U_0,$$

where $\tilde{U}_0, \tilde{U}_1$ *are subsets of* U_0 *and* U_1 *respectively.*

In the first (second) case the output 1 (0) of G remains always correct.

Theorem 3.3 can be proved by inspection for each of the considered gates and for every single stuck-at fault.

As an example we consider the *AND*-gate in Fig. 3.48 with the input lines 1, 2 and the output line 3. The sets U_0 and U_1 are $U_0 = \{00, 01, 10\}$ and $U_1 = \{11\}$.

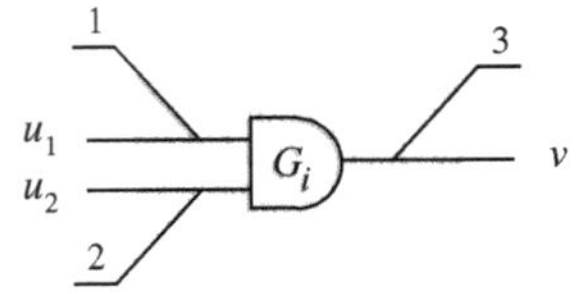

Fig. 3.48 *AND*-gate

The set of faults Φ consists of 6 single stuck-at faults

$$\Phi = \{1/1,\ 1/0,\ 2/1,\ 2/0,\ 3/1,\ 3/0\}$$

where $1/1$ denotes the stuck-at-1 fault at line 1.

For the following stuck-at faults we have:

- *stuck-at fault 1/0*:

$$\begin{aligned} 1 = G(u) \neq G_{1/0}(u) \text{ for } u \in \tilde{U}_1 = U_1 = \{11\} \\ 0 = G(u) = G_{1/0}(u) \text{ for } u \in U_0 = \{00, 01, 10\}. \end{aligned}$$

- *stuck-at fault 1/1*:

$$\begin{aligned} 0 = G(u) \neq G_{1/1}(u) \text{ for } u \in \tilde{U}_0 = \{01\} \subseteq U_{=}\{00, 01, 10\} \\ 1 = G(u) = G_{1/1}(u) \text{ for } u \in U_1 = \{11\}. \end{aligned}$$

- *stuck-at fault 3/1*:

$$\begin{aligned} 0 = G(u) \neq G_{3/1}(u) \text{ for } u \in \tilde{U}_0 = U_0 = \{00, 01, 10\} \\ 1 = G(u) = G_{3/1}(u) \text{ for } u \in \tilde{U}_1 = U_1 = \{11\}. \end{aligned}$$

In a similar way the remaining stuck-at faults of the *AND*-gate and also all single stuck-at faults of the other gates mentioned have to be considered. (In [77] a more general fault model of unidirectional gate faults is considered.)

Now the notion of a *self-dual fault-secure* circuit is introduced.

We consider a self-dual circuit $fc_{C_{sd}}$ implementing at its n outputs the n self-dual Boolean functions

$$\begin{aligned} y_1 &= f_1(x), \\ &\vdots \\ y_n &= f_n(x). \end{aligned}$$

The set of faults is denoted by Φ. In the presence of a fault $\varphi \in \Phi$ the faulty circuit implements the Boolean function $f_1(\varphi, x), \ldots, f_n(\varphi, x)$.

An error at an output y_j, $j \in \{1, \ldots, y_n\}$ will *not* be detected under the alternating inputs $x, \overline{x}$ if we simultaneously have

$$f_j(\varphi, x) \neq f_j(x) \quad \text{and} \quad f_j(\varphi, \overline{x}) \neq f_j(\overline{x}).$$

In this case in the presence of the fault φ the output y_j is erroneous under both the inputs x and $\overline{x}$ but it remains alternating.

If such a situation cannot occur, the output y_j is self-dual fault-secure.

Definition 3.14. The output y_j of a self-dual circuit $fc_{C_{sd}}$ with the input set X is self-dual fault-secure with respect to a fault $\varphi \in \Phi$ if for $x \in X$ either

$$f_j(\varphi, x) \neq f_j(x) \quad \textit{and} \quad f_j(\varphi, \overline{x}) = f_j(\overline{x}),$$

or

$$f_j(\varphi,x) = f_j(x) \quad and \quad f_j(\varphi,\overline{x}) \neq f_j(\overline{x}),$$

or

$$f_j(\varphi,x) = f_j(x) \quad and \quad f_j(\varphi,\overline{x}) = f_j(\overline{x}). \tag{3.34}$$

Definition 3.15. The output y_j is self-dual fault-secure if y_j is self-dual fault secure for every fault $\varphi \in \Phi$.

As the set of faults we consider here all single stuck-at faults.

Definition 3.16. A self-dual circuit fc_{sd} is self-dual fault-secure if every output of fc_{sd} is self-dual fault-secure.

Necessary and sufficient conditions for a self-dual circuit to be self-dual fault-secure are given in [77, 78].

By use of the following theorem an arbitrarily given combinational self-dual circuit can be transformed into a self-dual fault-secure circuit by a simple circuit transformation.

Theorem 3.4. *Let fc_{sd} be a self-dual circuit given as a netlist of M gates $G_1,\ldots,G_M$ and let y_j be an output of fc_{sd}.*

Then the output y_j of fc_{sd} is self-dual fault-secure with respect to all single stuck-at faults if for $i = 1,\ldots,M$ on all paths from the output of the gate G_i to the output y_j of fc_{sd} the parity or the modulo-2 sum of inverters is equal.

Proof:

If no fault occurs then

$$y_j(x) = f_j(x) \neq f_j(\overline{x}) = y_j(\overline{x})$$

for $x \in X$ is valid.

If a single stuck-at fault φ is only stimulated by x (and not by $\overline{x}$) we have

$$f_j(\overline{x}) = f(\varphi,\overline{x}).$$

The fault is detected if $f_j(x) \neq f(\varphi,x)$. For $f_j(x) = f(\varphi,x)$ no error occurs at the output y_j.

If φ is only stimulated by $\overline{x}$, the considerations are similar.

Let us now assume that both x and $\overline{x}$ stimulate φ. Let $G_i(x) = 0$. Then, according to Theorem 3.3 we have

$$G_i(\varphi,x) = G_i(\varphi,\overline{x}) = 1.$$

In the notation of the D-algorithm [79], a D (1 instead of 0) has to be propagated by x and by $\overline{x}$ to the output y_j.

If the parity or the modulo-2 sum of the inverters on all the paths from G_i to y_j is even, D is propagated to y_j, changing for x and also for $\overline{x}$ the correct output value 0 to the erroneous value 1. This is in contradiction to $y_j(x) \neq y_j(\overline{x})$ and it is impossible that simultaneously $f_j(x) \neq f_j(\varphi,x)$ and $f_j(\overline{x}) \neq f_j(\varphi,\overline{x})$ are valid. Either the fault will be detected or no error occurs at the output y_j. If the parity or the modulo-2 sum of the inverters on all the paths from G_i to y_j is odd, $\overline{D}$ is propagated to y_j, changing for x and also for $\overline{x}$ the correct output value 1 to the erroneous value 0, which is also in contradiction to $y_j(x) \neq y_j(\overline{x})$.

Now we describe a simple method for the transformation of an arbitrarily given self-dual circuit $f_{C_{sd}}$ into a self-dual fault-secure circuit $f_{C_{fs}}$. In the transformed self-dual fault-secure circuit $f_{C_{fs}}$ the parity of inverters on all paths from the output of a gate G_i to a circuit output y_j is equal.

As previously described we assume that $f_{C_{sd}}$ has the n outputs $y_1,\ldots,y_n$ and is given as a netlist of M gates $G_1,\ldots,G_M$.

The transformation of a given self-dual circuit into a self-dual fault-secure circuit consists of the following steps:

1. Every gate G_i is duplicated in the gates G_i^0 and G_1^1.
2. If the output of the gate G_k is directly connected to a circuit output y_l, we connect the output of G_k^0 to y_l. If the output of the gate G_k is connected to the circuit output y_l via an inverter, we connect the output of G_k^1 to y_l. (If, for instance, a *NOR*-gate is connected to a circuit output, we interpret this connection as a connection via an inverter).
3. If the output of a gate G_k is directly connected to an input of a gate G_l, then we connect the output of G_k^0 with the corresponding input of G_l^0 and the output of G_k^1 with the corresponding input of G_l^1. If the output of a gate G_k is connected via an inverter to an input of a gate G_l, then we connect the output of G_k^0 with the corresponding input of G_l^1 and the output of G_k^1 with the corresponding input of G_l^0.
4. All gates not connected to a circuit output are deleted (step by step).

The described circuit transformation is very similar to the circuit transformation already described in [19]. The transformation in [19] guarantees that every single stuck-at fault of the transformed circuit results in an unidirectional error at the circuit outputs, which can be detected by a Berger code checker.

Now we illustrate the described method using an example of a self-dual circuit shown in Fig. 3.49.

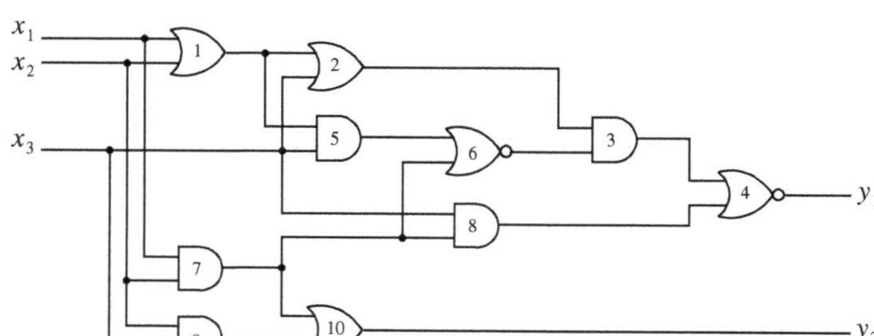

Fig. 3.49 Example of a self-dual circuit

The circuit consists of 10 gates numbered from 1 to 10. The only gates for which there are two different paths to a circuit output are the gates G_1 and G_7.

The paths p_1^1 and p_1^2 from G_1 to the output y_1 are

$$p_1^1 = G_1G_2G_3G_4 \quad \text{and} \quad p_1^2 = G_1G_5G_6G_3G_4.$$

The modulo-2 sum of inverters is 1 for p_1^1 and 0 for p_1^2.

The paths p_7^1 and p_7^2 from G_7 to the output y_1 are

$$p_7^1 = G_7G_6G_3G_4 \quad \text{and} \quad p_7^2 = G_7G_8G_4.$$

The modulo-2 sum of inverters is 0 for p_7^1 and 1 for p_7^2.

Now we duplicate for $i = 1,\dots,10$ every gate G_i of the original circuit of Fig. 3.49 in two gates G_i^0 and G_i^1 in Fig. 3.50.

Since in Fig. 3.49 the gate G_4 is connected to the output y_1 via an inverter, in Fig. 3.50 G_4^1 is connected to the output y_1 and G_4^0 is not connected to a circuit output.

In the original circuit G_3 is directly connected to G_4. Therefore in Fig. 3.50 G_3^0 is connected to G_4^0 and G_3^1 to G_4^1.

In Fig. 3.50 G_6^1 is connected to G_3^0 and G_6^0 is connected to G_3^1 since in Fig. 3.49 G_6 is connected via an inverter to to G_3.

The other duplicated gates shown in Fig. 3.50 are connected in a similar way.

The gates G_4^0, G_3^0, G_2^0, G_6^1, G_5^1, G_8^0, G_{10}^1, G_9^1 are not connected to any circuit output and removed.

The original non-self-dual fault-secure circuit of Fig. 3.49 consists of 10 gates compared to the 12 gates of the self-dual fault-secure circuit of Fig. 3.50.

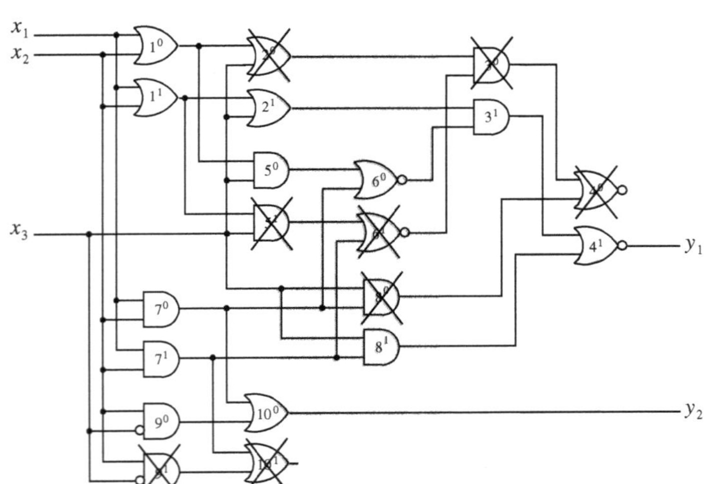

Fig. 3.50 Transformation of the example of Fig. 3.49 into a self-dual fault-secure circuit

The gates G_1 and G_7 are duplicated in G_1^0, G_1^1 and G_7^0, G_7^1 respectively.

In the transformed circuit of Fig. 3.50 there are single paths from all the duplicated gates G_1^0, G_1^1, G_7^0, G_7^1 to the circuit output y_1.

A similar transformation can be used for transforming an arbitrarily given combinational circuit into a circuit for which all single gate faults, including all single stuck-at faults, result in unidirectional faults at the circuit outputs [19].

In this section it was shown how the method of error detection using self-dual Boolean functions can be implemented.

Alternating inputs have to be applied to the self-dual circuit. Under alternating inputs the circuit outputs of these self-dual circuits are alternating. If, due to a fault in the self-dual circuit or due to *stuck-at faults at the input lines* the outputs are not alternating, a fault is detected.

Since for error detection always alternating inputs are applied, time redundancy is 100%, and the described method can be applied if time is not critical as, for instance, in mechanical control systems.

Two different methods for the transformation of an arbitrarily given combinational circuit in a self-dual circuit were described. It was assumed that the circuits were given as a netlist of gates.

In the first method (known for a long time) an additional input variable is added and the self-dual circuit is functional completely specified. The optimization of the self-dual circuit is reduced to the optimization of a completely specified Boolean function.

In the second method a complementary circuit is added to the original circuit such that the componentwise *XOR*-sums of the outputs of the original circuit and of the complementary circuit are arbitrary self-dual functions.

It was shown that for a given functional circuit there are many different possibilities to determine a complementary circuit that can be utilized for circuit optimization. This remains a challenging synthesis problem that has not yet been solved by the available synthesis tools. Some heuristical solutions based on the netlist of the original circuit were described in this section.

Error detection by self-dual parity and self-dual duplication was considered in more detail.

For error detection by ordinary duplication and comparison the original circuit and the duplicated circuit have to be separately implemented. This is different for self-dual duplication. Since the original circuit and the complementary circuit implement completely different functions, a joint implementation of both these circuits can be accomplished with a considerable reduction of the necessary area and a small reduction in the error detection probability.

Self-dual fault-secure circuits were considered at the end of this chapter. It was shown how a self-dual circuit, given as a netlist of gates, can be transformed into a self-dual fault-secure circuit by a simple circuit transformation.

It was mentioned that self-dual circuits can be operated in a fast mode without error detection and no reduction in speed, a slow error detection mode with alternating inputs and a self-test mode where only in the test mode alternating inputs are applied and no test responses had to be stored.

3.9 Error Detection with Soft Error Correction

In this section we show how error correction of soft errors, which are directly induced by radiation in the memory elements, and error detection by use of systematic codes for the errors caused by transient faults in the combinational parts of the circuit can be combined.

The rapidly shrinking dimensions of VLSI make it possible to design incredibly small chips. This trend of reduction in the size of transistors will continue at least in the near future.

However, because of the rapidly shrinking transistor sizes the number of transient faults in these small chips induced by radiation (not only in airplanes, but already at sea level), α-particles, crossover, electromagnetic fields and other reasons will dramatically increase in the near future.

A transient fault in the combinational part of the circuit will be reflected in the behavior of the system if an erroneous value is memorised in a latch or flip-flop of a register. This is only the case if logic, electrical and timing conditions are fulfilled [16].

The number of soft errors directly generated in the latches is about 20–40 times higher than the number of errors in the latches caused by transient faults in the combinational parts of the circuit [80].

Since the soft errors generated directly in the memory elements are so frequent in the evolving nano-technology of the near future, the applicability of traditional error detection methods will be limited. The resulting frequent error indications and the associated numerous interruptions of normal operation for frequent restarts of the system are intolerable for many applications.

To overcome this problem we propose in this section to combine fault tolerance for soft errors generated directly in the memory elements and error detection for the errors caused by transient faults in the combinational parts of the circuit.

The memory elements, and not the circuit itself, are duplicated and the soft errors directly induced in the memory elements are corrected by use of *C-elements*. Errors due to transient faults in the combinational part of the circuit are checked by error-detecting codes. The correction logic is included in the error detection.

3.9.1 Description of the Method

Now we describe how according to [81] the correction of soft errors in memory elements can be combined with error detection.

Soft errors in latches or flip-flops can be directly induced by radiation or caused by transient faults in the combinational part of the circuit.

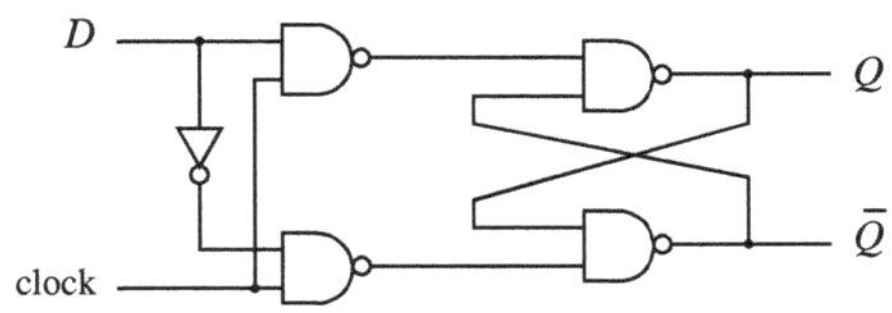

Fig. 3.51 D-latch

For error correction of soft errors directly induced in the memory elements the memory elements are duplicated and serially connected to *C-elements* [32]. All the other errors are checked by an error detection circuit.

Figure 3.51 shows a *D*-latch. If the clock signal *clock* = *1*, the value *D* of the data input determines the state of the latch and the latch is immune with respect to radiation-induced errors in its state.

If the clock signal *clock* = *0*, the state of the latch is disconnected from its input and is vulnerable to radiation-induced errors.

Thus we can expect that in the case of a soft error the state of the latch is correct in the first half of the clock cycle (for *clock* = *1*) and erroneous in the second half of the clock cycle (for *clock* = *0*).

If the latch is now duplicated, we can assume in the case of a soft error for *clock* = *1* that both of the duplicated latches are in the correct state and that for *clock* = *0* at most one of them is in an erroneous state.

To correct the erroneous value in one of the latches, the outputs of the duplicated latches are connected to a *C-element* with two inputs and one output [80].

A possible implementation of a *C-element* is shown in Fig. 3.52 with the input-output behavior given in Table 3.8.

If the two inputs y_i, y_i' are equal, $y_i = y_i' = y$ the output y_i^{corr} of the *C-element* is y. If the two inputs are not equal, $y_i \neq y_i'$, the *C-element* outputs its previous output value.

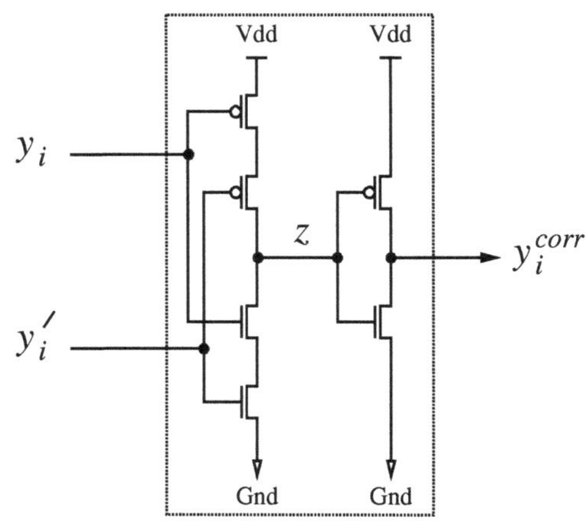

Fig. 3.52 Implementation of a *C-element*

Table 3.8 Input-output behavior of *C-element*

previous output	y_i	y'_i	y_i^{corr}
–	0	0	0
0	0	1	0
1	0	1	1
0	1	0	0
1	1	0	1
–	1	1	1

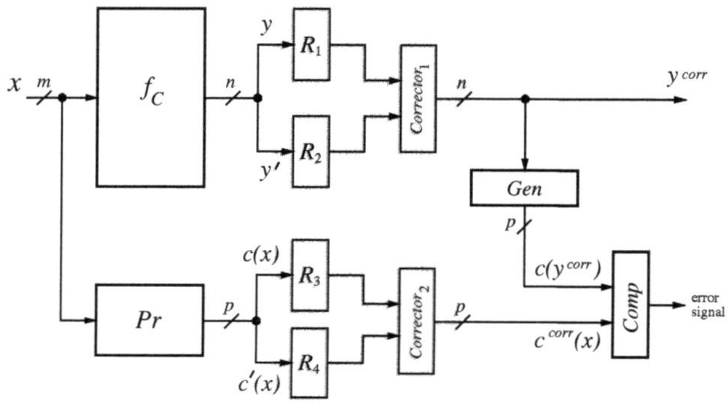

Fig. 3.53 Error correction combined with error detection for a systematic block code

Since in the case of a soft error in a latch the states of the duplicated latches in the first half of the clock cycle are correct and equal and since they are different in the second half of the clock cycle the soft error is corrected by the *C-element*.

Different implementations of *C-elements* are described for instance in [82, 83].

Figure 3.53 shows how error detection by systematic codes for transient faults in the combinational part of the circuit can be combined with error correction of soft errors in the latches or registers.

The combinational part f_C has m inputs $x = (x_1, \dots, x_m)$ and n outputs $y = (y_1, \dots, y_n)$ and implements at its outputs the n Boolean functions of length m

$$y_1 = f_1(x),$$
$$\vdots$$
$$y_n = f_n(x).$$

The outputs of f_C are duplicated in $y = y_1, \dots, y_n$ and $y' = y'_1, \dots, y'_n$ and captured in the two n-bit latches R_1 and R_2.

For $i = 1, \ldots, n$ the i-th components y_i, y'_i of the duplicated outputs of R_1 and R_2 are connected to a *C-element* C_i, correcting y_i, y'_i as y_i^{corr}. The corrector *Corrector1* consists of n *C-elements* $C_1, \ldots, C_n$.

The generator *Gen* determines the p check bits

$$c(y^{corr}) = c_1(y^{corr}), \ldots, c_p(y^{corr})$$

of the considered systematic block code from the n corrected information bits $y_1^{corr}, \ldots, y_n^{corr}$.

The predictor *Pr* determines the check bits $c(x) = c_1(x), \ldots, c_p(x)$ from the inputs variables $x = (x_1, \ldots, x_m)$.

The check bits $c(x) = c_1(x), \ldots, c_p(x)$ at the output of the predictor *Pr* are duplicated in

$$c(x) = c_1(x), \ldots, c_p(x)$$

and

$$c'(x) = c'_1(x), \ldots, c'_p(x).$$

The duplicated check bits $c(x)$ and $c'(x)$ are stored in the p-bit latches R_3 and R_4.

For $j = 1, \ldots, p$ the components c_j and c'_j of the latches R_3 and R_4 are connected to the j-th *C-element* C_j of the second corrector *Corrector2* and corrected as $c_j^{corr}(x)$.

The corrected check bits

$$c^{corr}(x) = c_1^{corr}(x), \ldots, c_p^{corr}(x)$$

and

$$c(y^{corr}) = c_1(y^{corr}), \ldots, c_p(y^{corr})$$

are compared by a (self-checking) comparator *Comp*.

Let us now assume that a radiation-induced error occurs in the i-th latch of R_1. Then for *clock* = *1* the states of both the i-th latches in R_1 and R_2 are equal and correct.

For *clock* = *0* the state of the i-th latch of R_1 is erroneous and different from the state of the i-th latch of R_2. The error is corrected by the i-th *C-element* C_i of the corrector *Corrector1*.

Similarly, all the other (multiple) soft errors in the latches are corrected either by the first corrector *Corrector1* or by the second corrector *Corrector2* as long as only one of the corresponding duplicated latches is faulty.

Let us now consider a transient fault in the combinational part f_C. A transient fault in the combinational part f_C may result in a soft error that changes the state of the i-th latches in both R_1 and R_2. Since the states of the i-th latches in both R_1 and R_2 are equal and erroneous they are not corrected by the i-th *C-element*.

If the error is detectable by the considered systematic code, the error is detected by comparing the check bits $c(y^{corr})$ and $c^{corr}(x)$.

Also transient faults in the predictor *Pr*, the correctors *Corrector1*, *Corrector2* and the generator *Gen* are detected if they are propagated to their outputs and latched.

Soft errors directly induced in the latches by radiation are about 20–40 times more frequent than soft errors resulting from transient faults in the combinational part.

If the soft errors directly induced in the registers would not be corrected, the functional mode of the circuit f_C would be interrupted too often by an error signal.

Correcting the soft errors frequently induced directly in the latches makes it possible to detect the transient faults generated in the combinational part of the circuit.

As an example of the proposed method we now consider error detection by a parity code as shown in Fig. 3.54.

The parity predictor *Pr* determines a single parity bit $P(x)$. The parity predictor *Pr* can be designed as a serial connection of the combinational circuit f_C and an XOR-tree, which is optimized by an available synthesis tool.

The parity bit $P(x)$ is duplicated in $P(x)$ and $P'(x)$ and stored in two latches L_3 and L_4.

The outputs of the latches L_3 and L_4 are connected to the corrector *Corrector2*, which consists of a single *C-element*. The corrected parity $P(x)$ is denoted by $P^{corr}(x)$.

The generator *Gen* is an XOR-tree generating at its outputs $P(y^{corr})$.

The comparator for comparing $P^{corr}(x)$ and $P(y^{corr})$ is a single XOR-gate. If a two-bit output is desired, no comparator is needed.

If the output of the parity predictor *Pr* is not duplicated and if the two latches L_3, L_4 and the connected *C-element* are replaced by a single latch L, the error correcting/detecting circuit is simplified. The simplified error correcting/detecting circuit for a parity code is presented in Fig. 3.55.

A soft error in the latch L in Fig. 3.55 will not be corrected but detected by comparing $P(x)$ and $P(y^{corr})$.

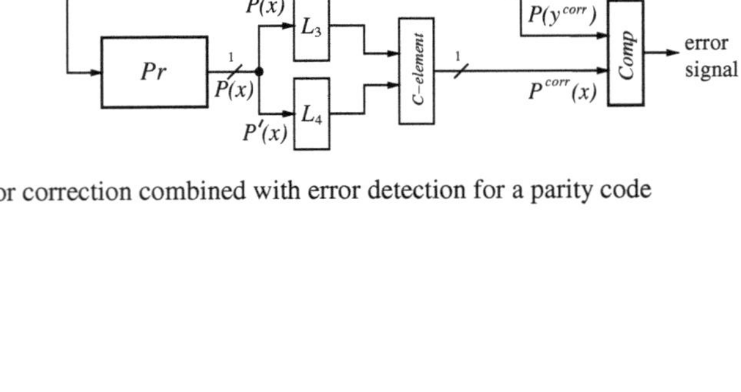

Fig. 3.54 Error correction combined with error detection for a parity code

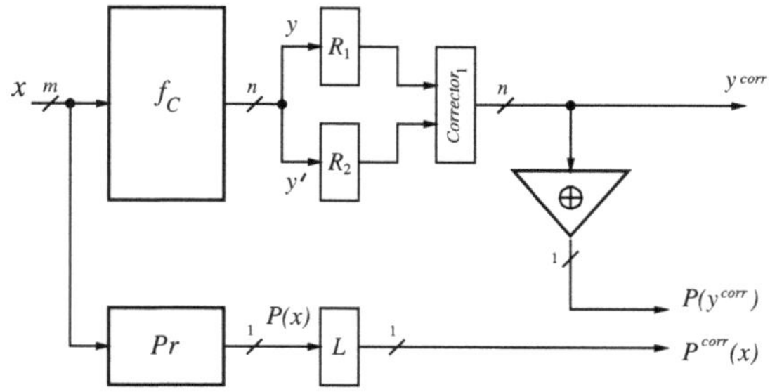

Fig. 3.55 Simplified error correction combined with error detection for a parity code

With a word length n of the n-bit latches R_1 and R_2 only 1 out of $2n+1$ soft errors will not be corrected but will be, as already pointed out, detected. Thus, for $n = 32$ only 1 out of 65 soft errors induced directly in the latches will not be corrected.

Using the described method of error detection with soft error correction for the most frequent errors, error detection will remain a useful method for circuits with a high soft-error rate of errors directly induced in the latches.

It was demonstrated in this section how error correction for soft errors directly induced in the memory elements can be beneficially combined with error detection for errors caused by transient faults in the combinational parts of the circuit by systematic codes.

Only the memory elements, and not the complete circuit, were duplicated for *correction*. Since soft errors can be expected to occur only in the second half of the cycle time in one of the duplicated memory elements they are corrected by two-input *C-elements*.

The number of soft errors directly induced in the latches is about 20–40 times higher than the number of errors in the latches caused by transient faults in the combinational part of the circuits. Therefore, most of the errors are corrected.

Only the relatively seldom errors due to transient faults in the combinational part have to be detected by an error detection circuit based on a systematic error detecting code. Restarts of the system are therefore also relatively seldom necessary. Errors in the correcting logic are also detected.

It was shown that the described method of error detection with soft error correction is a realistic possibility for the implementation of error detection for circuits with a high soft error rate.

Chapter 4
Concurrent Checking for the Adders

In this chapter it will be described how the general principles of concurrent checking can be applied for the design of self-checking adders.

Adders are examples of regular circuits built up by 1-bit adder cells. The internal structure of adders is well known and functionally they can be described by simple equations for the sum and carry bits. Therefore, the general methods for concurrent checking as presented in Chapter 3 of this book can be adapted to the internal structure of the different adder types. Also the adder cells and the internal structure of the adders can be modified to ease the design of error detection circuits.

Different possibilities of adaptation of the general error detection methods to self-checking adder designs, the modification of the internal structure for the computation of the sum and carry bits and the modification of the adder cells will be demonstrated in this chapter for the different adder types.

A high error detection probability for all errors caused by single stuck-at faults and for soft errors directly induced in the registers combined with a small area overhead and a short delay for the computation of the sum bits are the challenging design goals.

The most interesting concrete results described in this chapter are as follows:

- It will be demonstrated how the general method of code-disjoint partial duplication with parity checking for the non-duplicated part can be efficiently used for the design of self-checking adders. This will be shown for carry look-ahead adders, carry skip adders and carry select adders.
- A new type of adder, the sum bit-duplicated adder, will be introduced and it will be shown how this sum bit-duplicated adder can be used for the design of self-checking carry look-ahead and carry select adders with soft error detection in the output registers.

  It will be explained that by use of this sum bit-duplicated adder, almost the same error detection probability as duplication and comparison, but with a lower area overhead can be achieved for partially duplicated adders.
- It will be shown how the already existing functionally redundant parts of carry select adders and the carry look-ahead adders which were implemented to improve

the speed of the adders without error detection can be efficiently exploited to achieve a small additional area overhead for self-checking designs.

- It will be described how the area of a self-checking carry select adder can be reduced by replacing the duplicated adder blocks for the carry-in signals 1 by simple Add1-circuits.

The best possible state-of-the-art error detection circuits for the different adder types will be presented in this chapter.

4.1 Basic Types of Adders

In this section the different types of adders without concurrent checking will be described in brief.

Ripple adders, carry look-ahead adders, carry skip adders and carry select adders will be considered.

In a carry ripple adder the delay for the computation of the most significant sum bit is high. The most significant sum bit can only be computed if the carry signals of all the preceding adder cells have already been determined. It will be shown how this delay can be significantly reduced by use of a "fast ripple adder" for which the carry signals of the adder cells are split into two different carry signals and for which the delay is the delay of a single *NAND*-gate per adder cell.

In a carry look-ahead adder the carry-in signals of the adder cells are computed in a special look-ahead unit. The hierarchical structure will be explained for this look-ahead unit.

In a carry skip adder the carry-in signal of an adder block may skip the block if all the propagate signals of the adder cells of this block are equal to 1. Carry skip adders with constant and variable block sizes will be presented.

It will be explained how carry select adders with duplicated adder blocks for both the carry-in signals 0 and 1 are designed and that the adder blocks for the carry-in signal 1 can be replaced by much simpler Add1-circuits.

Addition is the most frequent and most important operation in digital computers. With an n-bit adder two binary-encoded n-bit operands $a = (a_0, \ldots, a_{n-1})$ and $b = (b_0, \ldots, b_{n-1})$ are added to form the n-bit sum $s = (s_0, \ldots, s_{n-1})$.

The least significant bits (LSB) of the operands and the sum are a_0, b_0 and s_0 respectively. The corresponding most significant bits (MSB) are a_{n-1}, b_{n-1} and s_{n-1}.

For an n-bit adder the $n+1$-bit carry vector c is denoted by $c = (c_{-1}, c_0, \ldots, c_{n-2}, c_{n-1})$ with the carry-in signal $c_{-1} = c_{in}$ and with the carry-out signal $c_{out} = c_{n-1}$.

For $i = 0, \ldots, n-1$ the i-th sum bit s_i and the i-th carry bit c_i are functionally determined as:

$$s_i = a_i \oplus b_i \oplus c_{i-1} \tag{4.1}$$

$$\begin{aligned} c_i &= a_i b_i \vee (a_i \vee b_i) c_{i-1} = \\ &= a_i b_i \vee (a_i \oplus b_i) c_{i-1}. \end{aligned} \tag{4.2}$$

The *generate* signal g_i and the *propagate* signal p_i are defined as

$$g_i = a_i b_i \quad (4.3)$$

$$p_i = a_i \oplus b_i. \quad (4.4)$$

A carry signal c_i will be generated in the i-th bit position from the operand bits a_i and b_i if the *generate* signal $g_i = a_i b_i$ is equal to 1. The *propagate* signal $p_i = a_i \oplus b_i$ determines whether the carry signal c_{i-1} of the $(i-1)$-th position will be propagated to c_i or not.

By use of the propagate and generate signals the i-th carry signal c_i can be expressed as:

$$c_i = g_i \vee p_i c_{i-1}. \quad (4.5)$$

A 1-bit full adder that implements the equations (4.1) and (4.2) is shown in Fig. 4.1.

For this adder the incoming carry $c_{i-1} = 1$ is not propagated to the outgoing carry c_i if $a_i = b_i = 1$. In this case we have $p_i = a_i \oplus b_i = 1 \oplus 1 = 0$. But for $a_i = b_i = 1$ the generate signal g_i is equal to 1, $g_i = a_i b_i = 1 \wedge 1 = 1$, and although $p_i = 0$, c_i remains correct.

In this section concurrent checking of the following types of adders is considered:

- carry ripple adders,
- carry look-ahead adders,
- carry skip adders
  and
- carry select adders.

First we briefly describe these four types of adders without error detection. Here we always assume that the word length of the operands and of the resulting sum is n. A more detailed description of different adders can be found, for instance, in [84, 85, 86, 87].

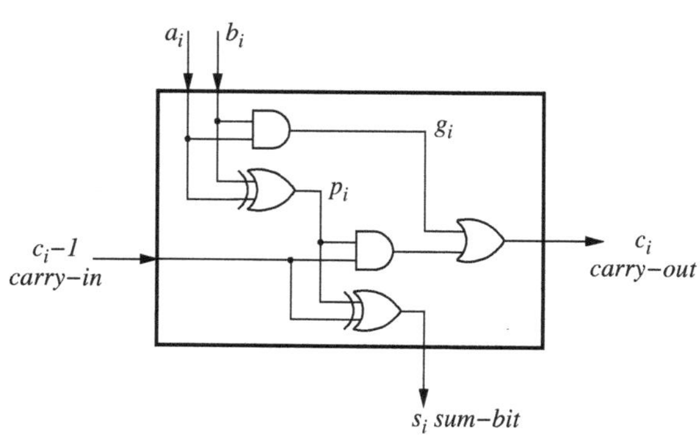

Fig. 4.1 1-bit full adder

In a carry ripple adder the delay for the computation of the most significant sum bit is high. The most significant sum bit can only be computed if the carry signals and the sum bits of all the preceding adder cells have already been determined. It will be shown how this delay can be significantly reduced by use of a "fast ripple adder" for which the carry signals of the adder cells are split into two different carry signals.

The hierarchical structure of the look-ahead unit will be explained for carry look-ahead adders.

Carry skip adders with constant and variable block sizes will be presented.

Traditional carry select adders with duplicated adder blocks for both the carry-in signals 0 and 1 and carry select adders where the blocks for the carry-in signal 1 are replaced by much simpler "Add1-circuits" will be described.

Carry Ripple Adder

The *carry ripple adder* is conceptually the simplest implementation of binary addition. A carry ripple adder consists of n 1-bit full adders cells A_i chained one after another as shown in Fig. 4.2.

For $i = 1,\dots,n-1$ the carry-in signal c_{i-1} of the i-th adder cell A_i is the carry-out signal of the preceding adder cell A_{i-1}.

As an example of adder cells the 1-bit full adder of Fig. 4.1 can be used. The number of adder cells corresponds to the length of the operands.

The main advantage of a carry ripple adder is its simplicity and the relatively low area needed for its implementation.

The main disadvantage of the carry ripple adder is the high delay for the computation of the most significant sum bit s_{n-1}. For $i = 0,\dots,n-1$ the sum bit s_i and the carry-out signal c_i can only be computed by the adder cell A_i when the carry-out signal c_{i-1} is already determined by the preceding adder cell A_{i-1}.

The carry propagation path (shown in Fig. 4.2 as a thick line) goes through all (n) adder cells.

The propagate signals p_i, $i = 0,\dots,n-1$, of all the adder cells are determined directly from the input operands a and b at the same time. In the first adder cell A_0 shown in Fig. 4.1 the carry-out bit c_0 is ready after three gate delays. In every adder cell A_i for the determination of c_i two gate delays are added and c_{n-2} which is needed for the computation of the most significant sum bit s_{n-1} is determined after

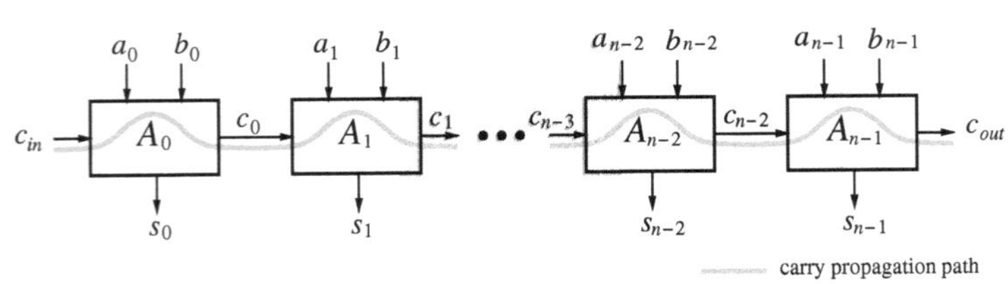

Fig. 4.2 Carry ripple adder

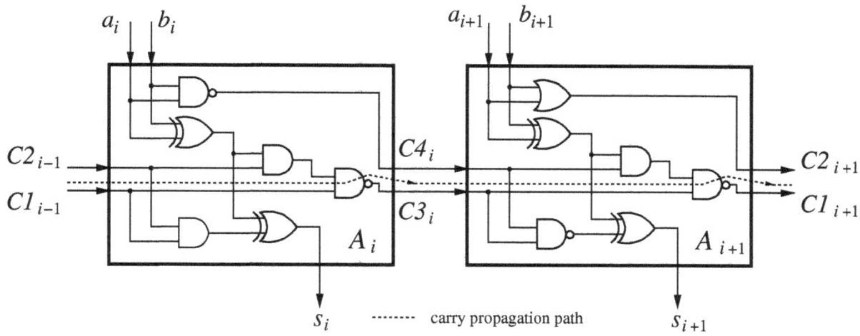

Fig. 4.3 "Fast" ripple adder cells

$3+(n-2)\times 2$ gate delays. The most significant sum bit s_{n-1} itself is computed by XORing the carry signal c_{n-2} with the propagate signal p_{n-1} of the last adder cell A_{n-1}.

To reduce the delay of a ripple adder two special adder cells as specified in [88] can be used. The resulting adder is called a *fast ripple adder*. Two successive adder cells of a *"fast" ripple adder* are shown in Fig. 4.3.

The main difference to a conventional adder is that the single carry signal c_j of a conventional ripple adder cell A_j is replaced by two carry signals $C1_j$ and $C2_j$ where j is even and by $C3_j$ and $C4_j$ where j is odd for a fast ripple adder:

$$c_j = \begin{cases} \overline{C1_j \wedge C2_j} & \text{for } j \text{ even} \\ C3_j \wedge C4_j & \text{for } j \text{ odd} \end{cases} \tag{4.6}$$

The adder is fast since the delay of the carry propagation is equal to the delay of a single NAND-gate per bit only.

Carry Look-Ahead Adder

In a *carry look-ahead adder*, as shown in Fig. 4.4, the carry-in signals for the adder cells and the carry-out signal of the n-bit adder are not determined by the corresponding preceding adder cells but by a special *carry look-ahead unit*. The adder cells $A_0,\ldots,A_{n-1}$ of the carry look-ahead adder calculate the sum bits $s_0,\ldots,s_{n-1}$, the propagate signals $p_0,\ldots,p_{n-1}$ and the generate signals $g_0,\ldots,g_{n-1}$ as inputs for the carry look-ahead unit. The carry look-ahead unit determines the carry-in signals $c_0,\ldots,c_{n-2}$ for the adder cells $A_1,\ldots,A_{n-1}$ and the carry-out signal c_{n-1} of the n-bit adder in accordance with the input carry $c_{-1}=c_{in}$, the propagate signals $p_0,\ldots,p_{n-1}$ and the generate signals $g_0,\ldots,g_{n-1}$ of the adder cells. The input carry c_{in} is already known before the addition.

The propagate signals $p_0,\ldots,p_{n-1}$ and the generate signals $g_0,\ldots,g_{n-1}$ are simultaneously determined in all the n adder cells $A_0,\ldots,A_{n-1}$ from the operands $a_0,\ldots,a_{n-1}$ and $b_0,\ldots,b_{n-1}$ with a single gate delay only.

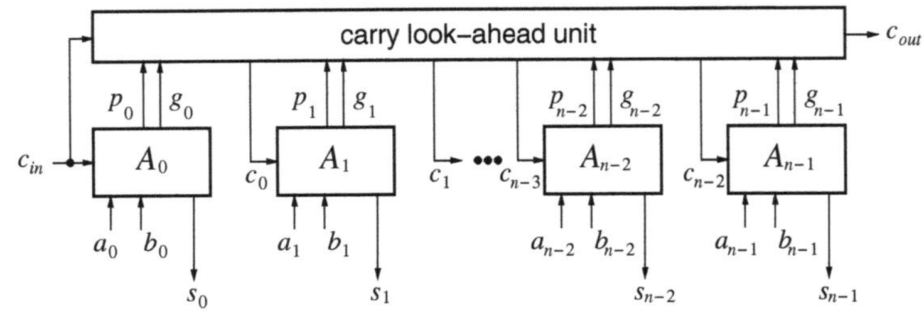

Fig. 4.4 Carry look-ahead adder

The carry look-ahead unit can be hierarchically designed by use of several levels of abstraction for the propagate and generate signals. The number of the levels depends on the word length n of the operands. We describe the carry look-ahead unit for the word lengths of 4, 16 and 64 bits.

For $n = 4$ the carries are determined as:

$$\begin{aligned} c_0 &= g_0 \vee p_0 c_{in} \\ c_1 &= g_1 \vee p_1 g_0 \vee p_1 p_0 c_{in} \\ c_2 &= g_2 \vee p_2 g_1 \vee p_2 p_1 g_0 \vee p_2 p_1 p_0 c_{in} \\ c_3 &= g_3 \vee p_3 g_2 \vee p_3 p_2 g_1 \vee p_3 p_2 p_1 g_0 \vee p_3 p_2 p_1 p_0 c_{in}. \end{aligned} \tag{4.7}$$

c_0 in 4.7 is directly determined by equation (4.5). For c_1 we have $c_1 = g_1 \vee p_1 c_0$ by equation (4.5) and with $c_0 = g_0 \vee p_0 c_{in}$ we conclude $c_1 = g_1 \vee p_1(g_0 \vee p_0 c_{in}) = g_1 \vee p_1 g_0 \vee p_1 p_0 c_{in}$. Similarly c_2 and c_3 are determined. Equation (4.7) describes a two-level implementation of the carry signals c_0, c_1, c_3 and c_4 in accordance with the propagate signals p_0, p_1, p_2, p_3, the generate signals g_0, g_1, g_2, g_3 and the carry-in signal c_{in} of the 4-bit adder. The carry signal c_3 is the carry-out signal c_{out} of the 4-bit carry look-ahead adder considered.

A carry look-ahead unit for a word length of 16 bits is shown in Fig. 4.5.

The 16-bit carry look-ahead unit is designed by use of 16 adder cells $A_0, \ldots, A_{15}$. The sixteen propagate signals $p_0, \ldots, p_{15}$ from these 16 adder cells A_0, $A_1, \ldots, A_{15}$

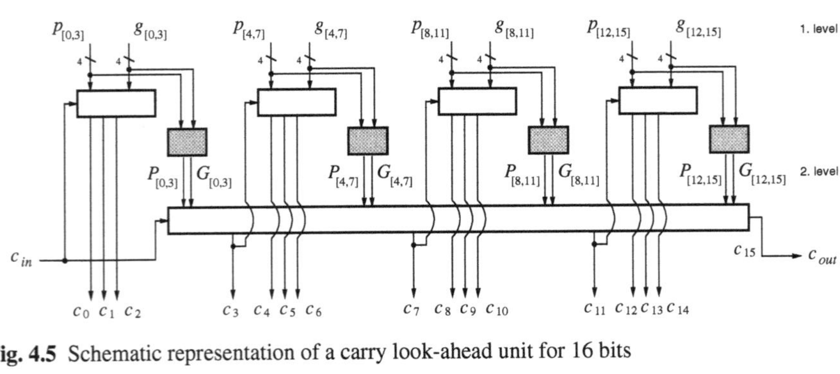

Fig. 4.5 Schematic representation of a carry look-ahead unit for 16 bits

are divided into four groups $[p_0,\ldots,p_3]$, $[p_4,\ldots,p_7]$, $[p_8,\ldots,p_{11}]$ and $[p_{12},\ldots,p_{15}]$. For each group a group-propagate signal $P_{[i,i+3]}$ at the second level of abstraction is determined as:

$$\begin{aligned} P_{[0,3]} &= p_0\ p_1\ p_2\ p_3 \\ P_{[4,7]} &= p_4\ p_5\ p_6\ p_7 \\ P_{[8,11]} &= p_8\ p_9\ p_{10}\ p_{11} \\ P_{[12,15]} &= p_{12}\ p_{13}\ p_{14}\ p_{15} \end{aligned} \tag{4.8}$$

Also from the sixteen generate signals $g_0,\ldots,g_{15}$ four group-generate signals $G_{[0,3]}$, $G_{[4,7]}$, $G_{[8,11]}$ and $G_{[12,15]}$ of the second level of abstraction are determined as:

$$\begin{aligned} G_{[0,3]} &= g_3 \vee p_3g_2 \vee p_3p_2g_1 \vee p_3p_2p_1g_0 \\ G_{[4,7]} &= g_7 \vee p_7g_6 \vee p_7p_6g_5 \vee p_7p_6p_5g_4 \\ G_{[8,11]} &= g_{11} \vee p_{11}g_{10} \vee p_{11}p_{10}g_9 \vee p_{11}p_{10}p_9g_8 \\ G_{[12,15]} &= g_{15} \vee p_{15}g_{14} \vee p_{15}p_{14}g_{13} \vee p_{15}p_{14}p_{13}g_{12} \end{aligned} \tag{4.9}$$

The shaded boxes in Fig. 4.5 represent the implementation of these group-propagate and group-generate signals.

The propagate and generate signals of the second level are used to determine the following carry signals:

$$\begin{aligned} c_3 &= G_{[0,3]} \vee P_{[0,3]}c_{in} \\ c_7 &= G_{[4,7]} \vee P_{[4,7]}G_{[0,3]} \vee P_{[4,7]}P_{[0,3]}c_{in} \\ c_{11} &= G_{[8,11]} \vee P_{[8,11]}G_{[4,7]} \vee P_{[8,11]}P_{[4,7]}G_{[0,3]} \vee P_{[8,11]}P_{[4,7]}P_{[0,3]}c_{in} \\ c_{15} &= G_{[11,15]} \vee P_{[11,15]}G_{[8,11]} \vee P_{[11,15]}P_{[8,11]}G_{[4,7]} \vee \\ &\qquad \vee P_{[11,15]}P_{[8,11]}P_{[4,7]}G_{[0,3]} \vee P_{[11,15]}P_{[8,11]}P_{[4,7]}P_{[0,3]}c_{in} \end{aligned} \tag{4.10}$$

The carries c_0, c_1, c_2 are implemented as in (4.7). Other groups of the carry signals $[c_4,c_5,c_6]$, $[c_8,c_9,c_{10}]$ and $[c_{12},c_{13},c_{14}]$ are determined in the same way, but the corresponding propagate and generate signals of the first abstraction level and the corresponding carries c_3 or c_7 or c_{11} or c_{in} are used. The carry c_{15} is the carry-out signal c_{out} of the 16-bit carry look-ahead adder.

For $n = 64$ a third level of the abstraction for the determination of the propagate and generate signals is used. The propagate and generate signals of this level are determined for the groups of four propagate and generate signals of the second level and used for the implementation of the carries c_{15}, c_{31}, c_{47} and c_{63}, where c_{63} is the carry-out signal c_{out} of the 64-bit carry look-ahead adder.

The look-ahead unit requires a large hardware overhead. However, the adder cells $A_i, i = 0,\ldots,n-1$ of the carry look-ahead adder are smaller than the adder cells in a carry ripple adder since no carry-out signal has to be implemented by the adder cells.

A good description of the structure of carry look-ahead adders can be found in [84]. Concrete implementations of the carry look-ahead unit are given, for instance, in [89].

Carry Skip Adder

In a carry skip adder the adder cells are divided into groups of equal or different size.

Figure 4.6 shows an n-bit carry skip adder.

In Fig. 4.6 the groups consisting of 4 adder cells each are represented as the blocks $B_j(4)$, $j = 1, \ldots, n/4$. Within every block $B_j(4)$ the adder cells are connected as in a carry ripple adder. The adder cells are usually implemented as 1-bit adders as shown for instance in Fig. 4.1. The "fast" ripple adder cells of Fig. 4.3 can also be used. If all the propagate signals of a block $B_j(4)$ are equal to 1, the carry-out signal of the previous block $B_{j-1}(4)$ can bypass or skip the block $B_j(4)$ to the next block $B_{j+1}(4)$ and the carry-out signal of the block $B_{j-1}(4)$ is the carry-in signal of the succeeding block $B_{j+1}(4)$.

A simple *skip* logic checks whether all the corresponding propagation signals are equal to 1. If this condition is not fulfilled, the carry-out signal of the last adder cell of the block $B_j(4)$ is the carry-in signal of the block $B_{j+1}(4)$. The long carry propagation path of a carry ripple adder can be accelerated by this simple technique.

The maximum delay of the carry skip adder is shown in Fig. 4.6 as the critical path. On the longest path the carry signal is generated in the least significant bit position (in the adder cell A_0). It ripples through the next three adder cells of the first block, skips over the blocks $B_2(4), \ldots, B_{n/4-1}(4)$ in the middle of the adder and ripples in the last block through the first three adder cells to the cell A_{n-1} to take part in the determination of the most significant sum bit s_{n-1}.

The delay of the carry skip adder is linearly dependent on the adder size n. However, this linear dependence is reduced by a factor of $1/k$.

The longest critical path of the carry skip adder can be minimized if blocks of variable sizes are used. The first and the last blocks are smaller than the intermediate blocks. Under the assumption that the 1-bit adder cell (according to Fig. 4.1) of the block has approximately the same delay as the skip logic [85], optimum variable block sizes can be predicted. For an n-bit carry skip adder these optimal variable block sizes are [85]:

$$\begin{aligned} k &\to (k+1) \to \ldots \to \\ &\to (k+t/2-1) \to (k+t/2-1) \to (k+t/2-2) \to \ldots \to \\ &\to (k+1) \to k, \end{aligned}$$

where $t = 2\sqrt{n}$ and k is the size of the smallest block.

The expected delay of the carry skip adder with variable block sizes is now a function of $\sqrt{n}$. To optimize the maximum delay in real designs the block sizes have to be experimentally determined. Starting from the expected values of the block

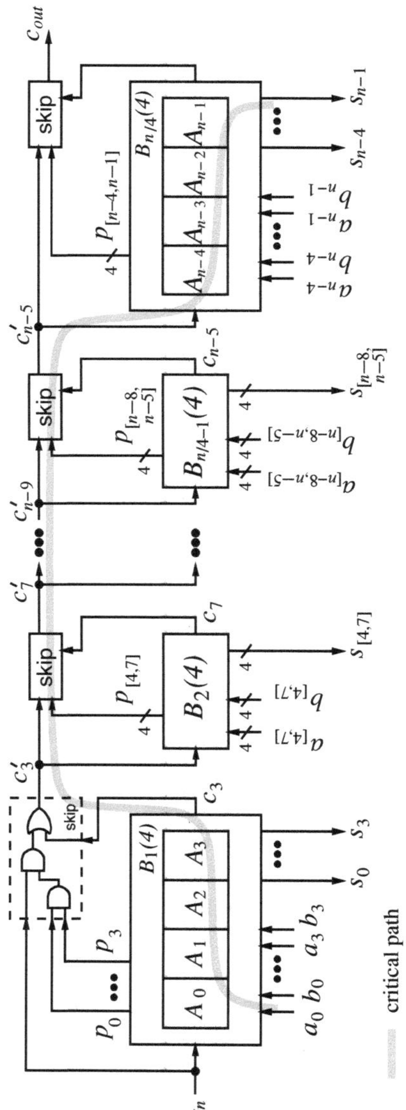

Fig. 4.6 Carry skip adder

sizes these block sizes have to be varied and the maximum delay is to be determined by use of a CAD synthesis tool.

Multi-level skip adders can also be used to further accelerate the addition [85].

Compared to the carry ripple adder the carry skip adder requires a larger implementation area. However, compared to a carry look-ahead adder the necessary area is relatively small.

For additional information about carry skip adders we recommend [85, 89, 87].

Carry Select Adder

Carry select adders are the fastest adders. The general structure of a carry select adder is shown in Fig. 4.7.

In the carry select adder the operands $a_0,\ldots,a_{n-1}$ and $b_0,\ldots,b_{n-1}$ are divided into groups of length k. In Fig. 4.7 the size k of every groups is 4 bits. The addition of the bits a_0, a_1, a_2, a_3 and b_0, b_1, b_2, b_3 of the first (least significant) group is performed by the adder block $B_1(4)$. The carry-in signal of this least significant adder block $B_1(4)$ is 0 (or in general the carry-in signal c_{in} of the carry select adder).

All the other 4-bit adder blocks are duplicated. In the adder blocks $B_j^0(4)$, $j = 2,\ldots,n/4$, the corresponding groups of operands are added with a carry-in signal 0 for all theses blocks. In the duplicated adder blocks $B_j^1(4)$, $j=2,\ldots,n/4$, the groups of operands are added with a carry-in signal 1 for all theses blocks. All the adder blocks compute their sum bits and carry-out signals in parallel.

The first block $B_1(4)$ implements the resulting sum bits $s_{[0,3]}$ and the carry-out signal c_3 of the first group directly.

Let us assume that the carry-out signal c_3 of the first adder block is equal to 0. This carry signal c_3 is used as the control signal of the multiplexors which select the sum bits $s^0_{[4,7]}$ and the carry-out signal c_7^0 of the block $B_2^0(4)$ with carry-in "0" as the resulting sum bits $s_{[4,7]}$ and the resulting carry-out signal c_7 of the second group. For $c_3 = 1$ the multiplexor selects the sum bits $s^1_{[4,7]}$ and the carry-out signal c_7^1 of the block $B_2^1(4)$ with carry-in "1". In a similar way, the carry-out signal of the previous group of adder blocks is the control signal of the corresponding multiplexors that select the resulting sum bits and the resulting carry-out bit of the next group.

A carry select adder is very fast. Its delay is determined by the delay for the implementation of the carry-out signal of the first adder block and the sum of the delays of the multiplexors in selecting the carry-out signals of the other adder blocks.

The delay of a carry select adder can be reduced if groups of variable block sizes are used. The first two (least significant) groups have to be of the same size. Then the control signal of the multiplexor, which is the carry-out signal of the first block, and both the carry-out signals of the duplicated second block as the data inputs of this multiplexor are available at the same time. While the selection of the carry-out signal of the second group is performed, the adder blocks of the next groups have some additional time (equal to the delay of a multiplexor) to compute more operand

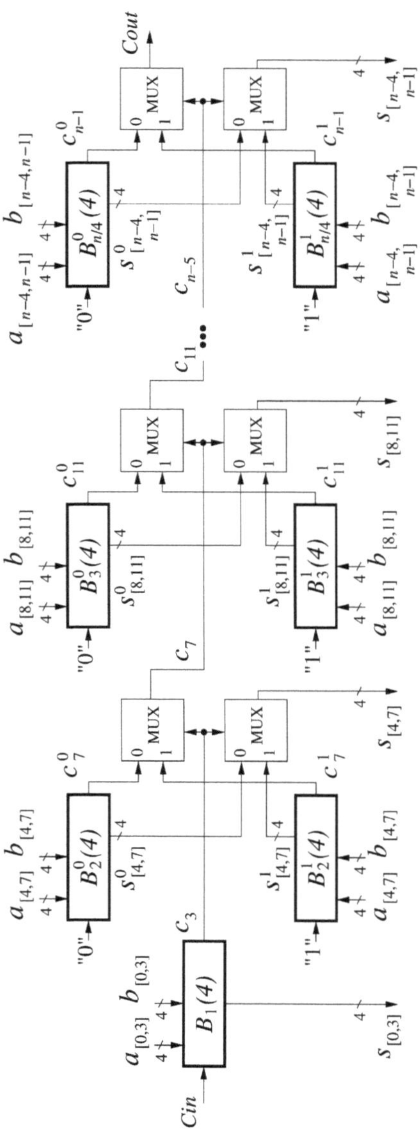

Fig. 4.7 Carry select adder

bits. Therefore, the size of these groups can be larger. The optimum sizes of the groups are dependent on the technology used and the library of the CAD tool.

Often the adder blocks are designed as carry ripple adders. But other adder types ("fast" ripple adder, carry look-ahead adder or carry skip adder) can also be used.

The necessary area overhead for the implementation of a carry select adder is very large since the adder blocks are duplicated and since additional multiplexors are used.

To reduce the necessary area overhead of the carry select adder according to [90] the duplicated adder blocks with the carry-in signals "1" can be replaced by much more simple *Add1-circuits*.

For the outputs of the adder blocks $B_j^0(4)$ (with carry-in "0") and $B_j^1(4)$ (with carry-in "1") of the cary-select adder in Fig. 4.7 we have for $j = 2, \ldots, n/4$,

$$s_4^1\, s_5^1\, s_6^1\, s_7^1\, c_7^1 = s_4^0\, s_5^0\, s_6^0\, s_7^0\, c_7^0 + 1,$$

$$s_8^1\, s_9^1\, s_{10}^1\, s_{11}^1\, c_{11}^1 = s_8^0\, s_9^0\, s_{10}^0\, s_{11}^0\, c_{11}^0 + 1,$$

$$\vdots$$

$$s_{n-4}^1\, s_{n-3}^1\, s_{n-2}^1\, s_{n-1}^1\, c_{n-1}^1 = s_{n-4}^0\, s_{n-3}^0\, s_{n-2}^0\, s_{n-1}^0\, c_{n-1}^0 + 1. \tag{4.11}$$

According to (4.11) the outputs of the adder blocks $B_j^0(4)$ (with carry-in "0") and $B_j^1(4)$ (with carry-in "1") $B_j^1(4)$ (with carry-in "1") arithmetically differ by 1 and the outputs of the adder block $B_j^1(4)$ (with carry-in "1") are obtained by adding a "1" to the outputs of the adder block $B_j^0(4)$ (with carry-in "0"). If a "1" is added to the outputs of the adder block $B_j^0(4)$ (with carry-in "0"), the least significant bits of these outputs are inverted from the least significant bit to the first "0".

Thus we have, for example:

$$\begin{array}{r} 1\,0\,1\,1\,0 \\ +\quad\quad\quad\;\; 1 \\ \hline \bar{1}\,\bar{0}\,1\,1\,0 = 0\,1\,1\,1\,0 \end{array} \quad \text{or} \quad \begin{array}{r} 1\,1\,1\,0\,1 \\ +\quad\quad\quad\;\; 1 \\ \hline \bar{1}\,\bar{1}\,\bar{1}\,\bar{0}\,1 = 0\,0\,0\,1\,1 \end{array}$$

The described modification of the outputs of the blocks $B_j^0(4)$ (with carry-in "0") is implemented by Add1-circuits. The adder blocks $B_j^1(4)$ (with carry-in "1") are replaced by Add1-circuits.

Of course the inputs of the Add1-circuit replacing the adder block $B_j^1(4)$ (with carry-in "1") are not the corresponding operand bits but the outputs of the adder block $B_j^0(4)$ (with carry-in "0").

The implementation of a block of 4 bits of a carry select adder by use of an Add1-circuit is shown in Fig. 4.8.

The Add1-circuit consists of a simple transistor logic block *TLB* and some XNOR-gates. The additional delay caused by the Add1-circuits is almost negligible.

Descriptions of carry select adders can be found, for instance, in [85, 86, 87, 89].

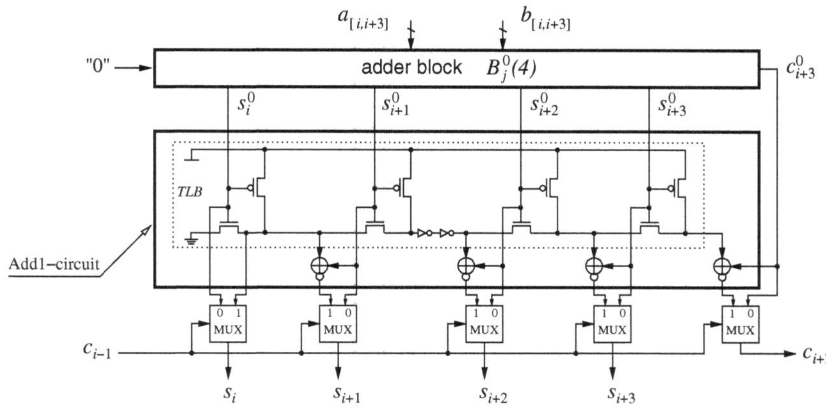

Fig. 4.8 Carry select adder by using Add1-circuit

In this section the structure of carry ripple adders, carry look-ahead adders, carry skip adders and carry select adders without concurrent checking were described.

It was shown that the large delay for the computation of the most significant sum bit of a carry ripple adder can be significantly reduced by use of a "fast ripple adder" for which the delay of the carry propagation per adder cell is only the delay of a single *NAND*-gate.

For carry look-ahead adders the hierarchical structure of the look-ahead unit was explained.

Carry skip adders with constant and variable block sizes were presented and it was discussed that the overhead for carry skip adders is between carry ripple adders and carry look-ahead adders.

Besides the traditional carry select adders with duplicated adder blocks for the two carry-in signals 0 and 1 carry select adders in which the adder blocks for the carry-in signal 1 are replaced by much simpler "Add1-circuits" were described.

4.2 Parity Checking for Adders

This section considers parity checking of adders. The basic equations for parity checking will be given. Single stuck-at faults may cause odd-bit and even-bit errors. It will be explained that odd-bit errors in the sum bits will be detected and that single-bit errors in the carry bits can be detected either if the carry bits are duplicated or if carry-dependent sum adder cells are used for the adder design.

Parity checking for adders was proposed in [91]. It is based on the equation

$$p_s = p_a \oplus p_b \oplus p_c, \tag{4.12}$$

where p_s, p_a, p_b and p_c are the parities of the sum s, the input operands a and b and the carries c respectively.

These parities are determined as

$$p_s = s_0 \oplus \ldots \oplus s_{n-1}, \tag{4.13}$$

$$p_a = a_0 \oplus \ldots \oplus a_{n-1}, \tag{4.14}$$

$$p_b = b_0 \oplus \ldots \oplus b_{n-1}, \tag{4.15}$$

$$p_c = c_{in} \oplus c_0 \oplus \ldots \oplus c_{n-2}. \tag{4.16}$$

The general structure of a parity-checked adder is shown in Fig. 4.9. The input operands $a = (a_0, \ldots, a_{n-1})$ and $b = (b_0, \ldots, b_{n-1})$ are supposed to be parity-encoded. The parities p_a and p_b of the input operands a and b are XORed to $p_a \oplus p_b$ and added modulo 2 with the parity p_c of the carries. The parity p_c of the carries is determined by an XOR-tree from the carry-in signals of all the adder cells. The modulo 2 sum $p_a \oplus p_b \oplus p_c$ of the parities p_a, p_b and p_c is compared with the parity p_s of the sum s. A difference indicates an error.

The necessary area overhead is relatively low. Two n-input XOR-trees and two XOR-gates are needed.

The parity-checked adder in Fig. 4.9 is *code-disjoint* [92] with respect to the parity code. All odd input errors are detected. Even input errors are not detected.

If the input operands are not parity-encoded, the corresponding parity-checked adder is not code-disjoint and input errors are not detected. In this case the parities p_a and p_b, have to be determined by two additional n-input XOR-trees as shown in Fig. 4.10.

The main problem of parity-checked adders is that single-bit errors in the carries could not be detected by parity checking as described till now. According to equation (4.1) a single-bit error in a carry simultaneously also causes an error in the next sum bit. Then the parities p_c and p_s are simultaneously erroneous, and the parity check according to equation (4.12) cannot detect this error.

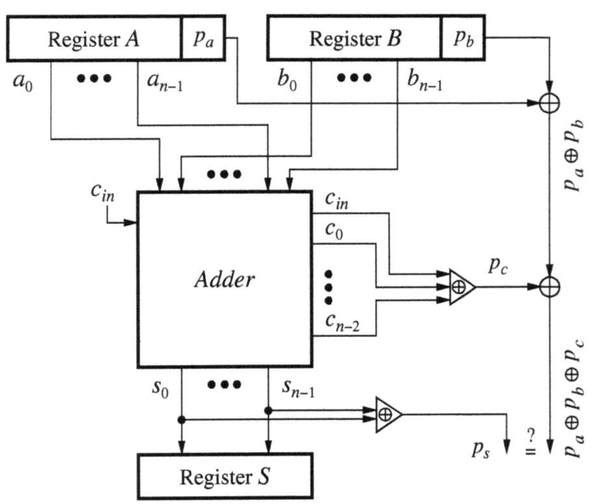

Fig. 4.9 Parity checking for adders

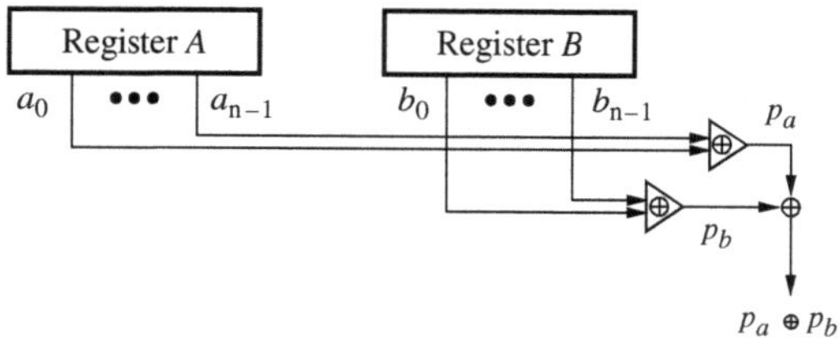

Fig. 4.10 Determination of the parities p_a and p_b for the non-parity-encoded input operands a and b

Up until now there were two known solutions to also detect single-bit errors in the carries. These are duplication and comparison of the carries and the design of carry-dependent sum adders.

1. Carry duplication: The hardware for the generation of the carries is duplicated. Errors in the carries are detected by comparing the duplicated carries by a (self-checking) comparator.
2. Carry-dependent sum adder: As proposed in [93, 91] special *carry-dependent sum adder* cells are used. A single error in a carry signal always results in an odd number of erroneous sum and carry bits in a carry-dependent sum adder, which are detected by parity checking according to equation (4.12).

In a carry-dependent sum adder the sum bit s_i is determined by use of a special auxiliary function f_i as:

$$s_i = f_i \oplus c_i \tag{4.17}$$

with

$$\begin{aligned} f_i &= \overline{a_i b_i c_{i-1} \vee \overline{a}_i \overline{b}_i \overline{c}_{i-1}} = \\ &= (a_i \oplus b_i) \vee (a_i \oplus c_{i-1}) = \\ &= (a_i \oplus b_i) \vee (b_i \oplus c_{i-1}). \end{aligned} \tag{4.18}$$

The dependencies between f_i and c_i with respect to an erroneous carry-in signal c_{i-1} are expressed by

Lemma 1 *If the carry-in c_{i-1} of the adder cell A_i is erroneous, then either the auxiliary function f_i or the carry-out c_i of the adder cell A_i is in error but never both.*

Lemma 1 is true due to the functional dependencies between f_i, a_i, b_i, and c_i and it is independent of the kind of implementation of the adder cell A_i. This can be seen from Table 4.1. If c_{i-1} is erroneously changed into $\overline{c}_{i-1}$, either $f_i(a_i, b_i, \overline{c}_{i-1})$ is different from $f_i(a_i, b_i, c_{i-1})$ or $c_i(a_i, b_i, \overline{c}_{i-1})$ is different from $c_i(a_i, b_i, c_{i-1})$ but never both. The values of $f_i(a_i, b_i, \overline{c}_{i-1})$, which are different from $f_i(a_i, b_i, c_{i-1})$, and the values $c_i(a_i, b_i, \overline{c}_{i-1})$, which are different from $c_i(a_i, b_i, c_{i-1})$, are underscored in Table 4.1.

Table 4.1 Functional dependencies of f_i and c_i in a carry-dependent sum adder for error-free carry-in c_{i-1} and for erroneous carry-in $\overline{c}_{i-1}$

$a_i\ b_i$	c_{i-1}	$\overline{c}_{i-1}$	$f_i(a_i,b_i,c_{i-1})$	$f_i(a_i,b_i,\overline{c}_{i-1})$	$c_i(a_i,b_i,c_{i-1})$	$c_i(a_i,b_i,\overline{c}_{i-1})$
0 0	0	1	0	**1**	0	0
0 0	1	0	1	**0**	0	0
0 1	0	1	1	1	0	**1**
0 1	1	0	1	1	1	**0**
1 0	0	1	1	1	0	**1**
1 0	1	0	1	1	1	**0**
1 1	0	1	1	**0**	1	1
1 1	1	0	0	**1**	1	1

In this section parity checking of adders was described. The basic equations for parity checking were given and it was explained that odd sum bit errors and also single errors of the carry bits are detected if either the carry bits are duplicated or if carry-dependent sum adder cells were used for the adder design.

4.3 Self-Checking Adders

In this section self-checking carry look-ahead adders, carry skip adders and carry select adders will be described.

4.3.1 Self-Checking Carry Look-Ahead Adders

A parity-checked carry look-ahead adder with duplicated carries, a parity-checked carry look-ahead adder with carry-dependent sum adder cells and a sum bit-duplicated carry look-ahead adder will be described. The inputs are supposed to be parity-encoded and it will be shown that all the presented designs are code-disjoint with respect to a parity code.

- *Parity-checked carry look-ahead adder with duplicated carries*
  Unlike a carry look-ahead adder without error detection, inverted carries are also generated in the adder cells. These inverted carries will be compared with the corresponding carries generated by the look-ahead unit by means of a self-checking two-rail checker. The original parity p_c and the inverted parity of the carries $\overline{p}_c$ are determined by this two-rail checker at its outputs. The input parities of the operands are XORed with the parity of the carries and this *XOR*-sum will be compared with the parity of the sum bits for error detection.
- *Parity-checked carry look-ahead adder with carry-dependent sum adder cells*
  Instead of duplicating all the carries in the adder cells in this design carry-dependent sum adder cells are used. The property that for a carry-dependent

sum adder either the sum bit or both the carry-out and the sum bit of a carry-dependent sum adder cell are simultaneously erroneous is utilized in the design. The adder cells are divided into groups connected as ripple adders. Only for the inputs of these groups the (partial) look-ahead unit determines the corresponding carries, which are compared with the carry-out signals of the preceding groups. For larger groups the number of carries which have to be generated by the partial look-ahead unit is smaller and the necessary area will decrease but with growing group sizes the adder will become slower. On the contrary, for smaller groups of adder cells the delay for the computation of the most significant sum bit will be reduced, but the necessary area increases.

- *Sum bit-duplicated carry-look ahead adder*
  The sum bit-duplicated adder is an example of error detection by partial duplication. The non-duplicated part of the combinational circuit is concurrently checked by parity prediction. The sum bits of the adder cells are duplicated in inverse form and stored in duplicated output registers. The inputs of the i-th adder cell A_i are the corresponding bits a_i and b_i of the operands and both the carries c_i and c_{i-1}, which are supplied by the carry look-ahead unit. The main advantage of the sum bit-duplicated adder is that all soft errors (even or odd) directly induced in the output registers are detected, and that the error detection capability of the sum bit-duplicated carry look-ahead adder is almost the same as for duplication and comparison but with a lower area overhead.

Parity-Checked Carry Look-Ahead Adder with Duplicated Carries

The first design is based on parity checking with duplicated carries as introduced in [94]. The structure of the parity-checked carry look-ahead adder according to [94] is shown in Fig. 4.11.

The input operands $a = (a_0, \ldots, a_{n-1})$ and $b = (b_0, \ldots, b_{n-1})$ are stored in the registers A and B respectively and supposed to be parity-encoded. The corresponding parity bits of the operands are denoted by p_a and p_b respectively. The carry signals are duplicated to also detect single-bit errors in the carries. The carry look-ahead unit generates the carry signals for the adder cells $A_0, \ldots, A_{n-1}$, and, differently to a carry look-ahead adder without error detection, the adder cells $A_0, \ldots, A_{n-2}$ implement the inverted carries additionally to the sum bits. The most significant adder cell A_{n-1} implements both the inverted carry $\overline{c}_{n-1}$ and the non-inverted carry c_{n-1}. The input carry of the adder itself is also duplicated in c_{in}^1 and c_{in}^2.

The pairs of the duplicated carries $[\overline{c}_{in}^1\ c_{in}^2]$, $[\overline{c}_0\ c_0], \ldots, [\overline{c}_{n-2}\ c_{n-2}]$ are compared by a self-checking two-rail checker *TRC*. In [94] the self-checking two-rail checker according to [27] is used.

This self-checking two-rail checker has the property that at its two outputs the parity p_c and the inverted parity $\overline{p}_c$ of its inputs are implemented, and no additional XOR-tree is needed to determine the parity p_c of the internal carries of the adder cells.

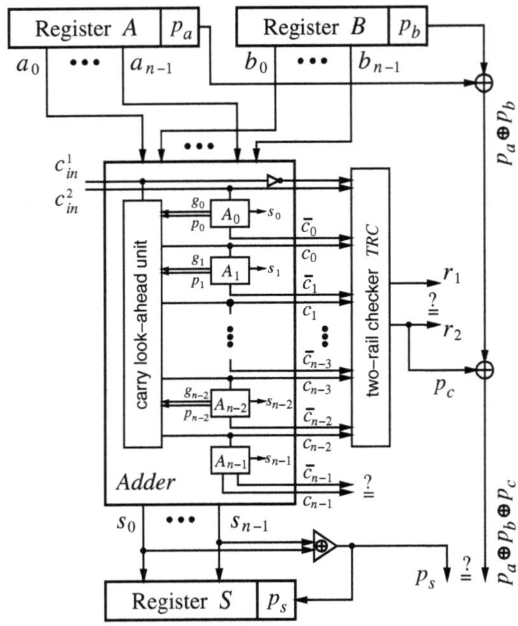

Fig. 4.11 Parity-checked carry look-ahead adder with duplicated carry signals

In accordance with equation (4.12), the parities p_a, p_b and p_c are XORed and compared with the parity p_s of the sum bits $s = (s_0, \ldots, s_{n-1})$, which is implemented by an XOR-tree.

The carry look-ahead unit in the parity-checked carry look-ahead adder with the duplicated carries is the same as for an ordinary carry look-ahead adder as described in Section 4.1.

The adder cells A_i for $i = 0, \ldots, n-2$ and A_{n-1} are shown in Fig. 4.12. For $i = 0, \ldots, n-2$ the adder cell A_i implements (Fig. 4.12a) the propagate signal p_i and generate g_i signals ($i = 0, \ldots, n-2$) as inputs for the carry look-ahead unit, the sum bit s_i and the inverted carry-out $\overline{c}_i$. The propagate signal p_i is also used in the adder cell for the generation of the sum bit s_i and the inverted carry-out bit $\overline{c}_i$. The generate signal g_i is not shared with any other signal. The most significant adder cell A_{n-1} (Fig. 4.12b) implements the sum bit s_{n-1}, the carry-out bit c_{n-1} and the inverted carry-out bit $\overline{c}_{n-1}$. Also in this adder cell the propagate signal p_{n-1} is used for the determination of the sum bit s_{n-1} and both the inverted and non-inverted carry-out bits. The carry-out bit c_{n-1} is the output carry of the adder.

The parity-checked carry look-ahead adder with duplicated carries detects the following errors:

- Odd errors in the input operands:
  An input error changes the correct input operands a and b into the erroneous operands a' and b'. Since the error is assumed to be odd, we have $p_a \oplus p_b \neq$

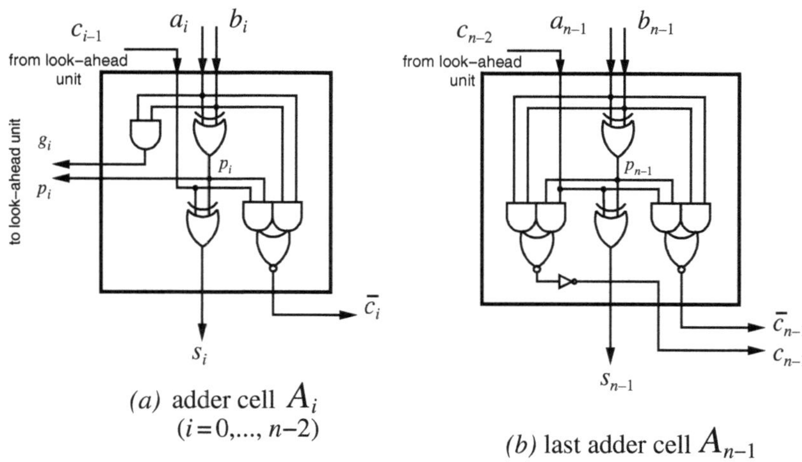

Fig. 4.12 Implementation of the cells of the parity-checked carry look-ahead adder in Fig. 4.11

$p_{a'} \oplus p_{b'}$. The sum bits $s'_0, \ldots, s'_{n-1}$ and the carry-bits $c'_0, \ldots, c'_{n-1}$ and the inverted carry bits $\overline{c}'_0, \ldots, \overline{c}'_{n-1}$ are determined for the erroneous operands a' and b'. We have $p'_s = p_{a'} \oplus p_{b'} \oplus p_{c'} \neq p_a \oplus p_b \oplus p_{c'}$, and the error is detected and the adder is code-disjoint.

- All stuck-at faults in the carry look-ahead unit:
  These faults change the internal carry signals $c_0, \ldots, c_{n-2}$ generated by the look-ahead unit and they are detected by comparing the carry pairs $[\overline{c}_0\ c_0], \ldots,$ $[\overline{c}_{n-2}\ c_{n-2}]$ with the two-rail checker *TRC*.
- All single stuck-at faults in the adder cells:
  - faults in an adder cell A_i of Fig. 4.12a
    Only faults in the XOR-gate generating the propagate signal p_i can result in multiple errors at the outputs p_i, s_i and $\overline{c}_i$.
    Let us assume that p_i is erroneous. Since we have $s_i = a_i \oplus p_i$ and since the operation $\oplus$ is uniquely invertible also the sum bit s_i is simultaneously erroneous. The carry signal c_i and the inverted carry signal $\overline{c}_i$ are functionally determined by the erroneous propagation signal p_i, the correct operand bits a_i, b_i and the incoming carry bit c_i by the carry look-ahead unit and in the adder cell A_i. They are either both erroneous or both correct. If the carry signals are both correct, only the sum bit s_i is erroneous and this error will be detected by parity checking.
    If both the carries c_i and $\overline{c}_i$ are erroneous, the sum bit $s_{i+1} = a_{i+1} \oplus b_{i+1} \oplus c_i$ is also erroneous. Since now s_i, s_{i+1} and c_i are simultaneously erroneous the error is detected by parity checking.
    Every other single stuck-at fault can result in a single-bit error of one of the output signals g_i, s_i or $\overline{c}_i$ only.
    Errors of c_i or $\overline{c}_i$ are detected by comparing the carry signals generated by the look-ahead unit and the inverted carry signal generated by the adder cells.

Errors of s_i are detected by parity checking and errors of g_i if they for the first time result in an erroneous carry signal of the look-ahead unit. Then this erroneous carry signal generated by the look-ahead unit is compared with the correct inverted carry signal from the corresponding adder cell.

- faults in the most significant adder cell A_{n-1} of Fig. 4.12b
  Similarly to the adder cell A_i of Fig. 4.12a only faults in the XOR-gate generating the propagate signal p_{n-1} can result in multiple errors at the outputs p_{n+1}, s_{n+1}, $\overline{c}_{n+1}$ and c_{n+1}. Stuck-at faults at the inputs/output of the XOR-gate, which implements the propagate signal p_{n-1}, always results in an erroneous sum bit s_{n-1}. The erroneous sum bit s_{n-1} changes the parity p_s of the sum. The carry-out bit c_{n-1} and the inverted carry-out bit $\overline{c}_{n-1}$ are not used for the determination of the parity p_c of the internal carries and do not influence p_c (see Eq. 4.16). Therefore, the stuck-at faults at the inputs/output of the XOR-gate, which implements the propagate signal p_{n-1}, are detected by the comparison of p_s with $p_a \oplus p_b \oplus p_c$. This comparison also checks the XOR-gate for the implementation of the sum bit s_{n-1}.
  The stuck-at faults at the inputs/output of the gates for the implementation of the carry-out bit c_{n-1} and for the implementation of the inverted carry-out bit $\overline{c}_{n-1}$ are detected by the comparison of c_{n-1} with $\overline{c}_{n-1}$.

- all single stuck-at faults in the XOR-tree for the implementation of p_s and of the two XOR-gates for the implementation of $p_a \oplus p_b \oplus p_c$ are detected by comparing p_s with $p_a \oplus p_b \oplus p_c$.
- all single stuck-at faults of the two-rail checker *TRC* are detected since the two-rail checker according to [27] is self-checking.
- odd errors in the sum bit register *S* are detected
  Since the register *S* is parity-protected only odd errors can be detected.

Compared to a carry look-ahead adder without error detection the additional hardware for a parity-checked self-checking carry look-ahead adder consists of the following:

- Two XOR-trees for determination of the input parities p_a and p_b of the operands a and b if the inputs are not yet parity-encoded,
- $3 \times (n+1)$ additional two-input gates for the implementation of the inverted carry signals in each of the n adder cells,
- a self-checking two-rail checker to compare the duplicated carries (two n-bit words have to be compared),
- an XOR-tree for the implementation of the parity p_s of the n sum bits,
- two XOR-gates for the implementation of the XOR-sum of the paritites p_a, p_b and p_c.

The propagate signals are simultaneously used for the implementation of the sum bits and the inverted carry-out bits in the adder cells. Because of this additional fan-out the computation of the sum s in the parity-checked carry look-ahead adder can be expected to be somewhat delayed compared to a carry look-ahead adder without error detection. But this delay is negligible. Parity checking and the comparison of the duplicated carries by the two-rail checker are performed with some latency compared to the calculation of the sum s.

Parity-Checked Carry Look-Ahead Adder with Carry-Dependent Sum Adder Cells

The next design is also based on parity checking, but instead of duplicated carries carry-dependent sum adder cells are used as proposed in [95]. The structure of the parity-checked carry look-ahead adder according to [95] is shown in Fig. 4.13.

The input operands $a = (a_0, \ldots, a_{n-1})$ and $b = (b_0, \ldots, b_{n-1})$ are again parity-encoded with the corresponding parity bits p_a and p_b. The adder cells are divided into groups of adder cells which are serially connected to carry ripple adders. In Fig. 4.13 every group of these adder cells consists of four carry-dependent sum adder cells. The duplicated carry-in signal c_{in}^1 of the adder is the carry-in signal for the adder cell A_0 of the first group. The carry-in signals for other groups of the adder cells are implemented by a partial carry look-ahead unit. The partial carry look-ahead unit determines not every carry signal, but only every fourth carry c'_3, $c'_7, \ldots, c'_{n-5}$, c'_{n-1}. The inputs of the partial carry look-ahead unit are the duplicated carry-in signal c_{in}^2 of the adder and for $i = 0, \ldots, n-1$ the propagate signals p_i and the generate signals g_i of the adder cells A_i.

Every adder cell A_i also implements the sum bit s_i and the carry-out bit c_i.

The carries c'_3, $c'_7, \ldots, c'_{n-5}$, c'_{n-1} implemented by the partial carry look-ahead unit and the carries c_3, $c_7, \ldots, c_{n-5}$, c_{n-1} implemented by the adder cells A_3, $A_7, \ldots, A_{n-5}$, A_{n-1} are compared by a self-checking two-rail checker TRC according to [27]. One of the outputs, say r_2, of the two-rail checker TRC implements the parity of the carries c_3, $c_7, \ldots, c_{n-5}$. The comparison of the most significant carries is separately implemented. The output r_2 of the two-rail checker TRC is added modulo 2 with the other non-duplicated carries to form the parity p_c of the internal carries of the adder cells. The parities p_a, p_b and p_c are XORed and compared with the parity p_s of the sum bits $s = (s_0, \ldots, s_{n-1})$. The parity p_s is implemented by an XOR-tree.

A possible implementation of a carry-dependent sum adder cell A_i of the described parity-checked carry look-ahead adder is given in Fig. 4.14. The adder cell A_i implements the propagate signal p_i, the generate signal g_i for the partial carry look-ahead unit, the carry-out bit c_i and the sum bit s_i ($s_i = f_i \oplus c_i$). The adder cell consists of two separate subcircuits – one for the implementation of p_i, g_i and f_i and one for the implementation of c_i. Every one of these subcircuits can be separately optimized to reduce the necessary area overhead. A joint implementation of these subcircuits can also be considered, but it may reduce the detectability of faults. (A special joint implementation is used for the design of self-checking multipliers).

The adder cells A_3, $A_7, \ldots, A_{n-5}$,A_{n-1} are the last adder cells in the groups. The carry-out signals c_3, $c_7, \ldots, c_{n-5}$, c_{n-1} of these adder cells are not used as carry-in

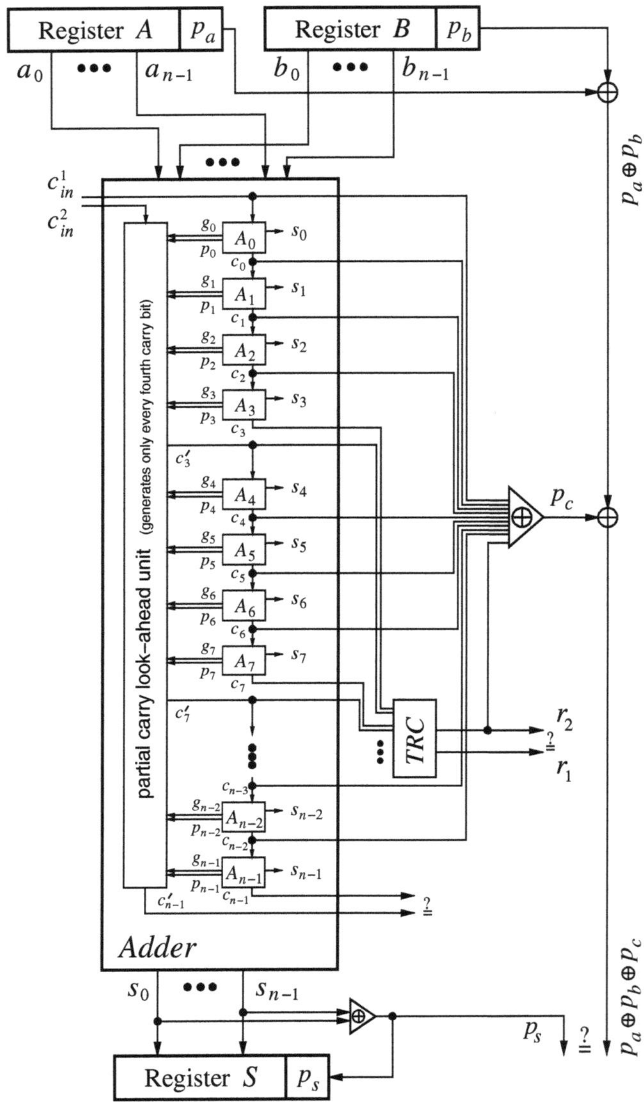

Fig. 4.13 Parity-checked carry look-ahead adder with carry-dependent sum adder cells

signals for the next adder cells. These carry-out signals are compared with the carry-out signals of the partial carry look-ahead unit. Therefore, the adder cells A_3, $A_7, \ldots, A_{n-5}, A_{n-1}$ can also be implemented as shown in Fig. 4.12a. The number of adder cells in a group can be chosen arbitrarily. For larger groups the number of the carries implemented by the partial carry look-ahead unit is smaller and the area overhead can be reduced. But in this case the delay for the calculation of the sum is increased. For smaller groups of adder cells the delay is reduced, but the area required increases.

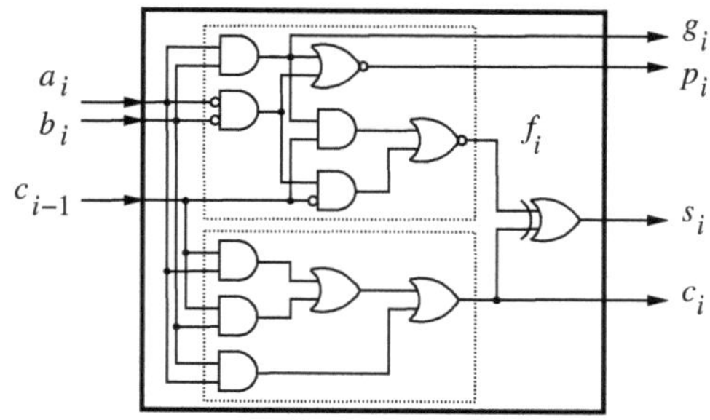

Fig. 4.14 Implementation of the carry-dependent sum adder cell for the parity-checked carry look-ahead adder in Fig. 4.13

If the last adder cells in the groups are implemented as shown in Fig. 4.12 and if the number of the adder cells in the groups is equal to one, the described solution for the self-checking carry look-ahead adder is equivalent to a self-checking carry look-ahead adder with duplicated carries.

The parity-checked carry look-ahead adder with carry-dependent sum adder cells detects the following errors:

- Odd errors in the input operands:
  The explanation is the same as for the parity-checked carry look-ahead adder with duplicated carries.
- All stuck-at faults in the partial carry look-ahead unit:
  These faults change the carry signals $c'_3, c'_7, \ldots, c'_{n-5}, c'_{n-1}$ and they are detected by the comparison of the carry pairs $[c'_3\ c_3], [c'_7\ c_7], \ldots, [c'_{n-5}\ c_{n-5}], [c'_{n-1}\ c_{n-1}]$ by the two-rail-checker *TRC*.
- All single stuck-at faults in the adder cell A_i of Fig. 4.14:
  The adder cell A_i in Fig. 4.14 consists of two separate subcircuits. One of them implements the propagate p_i and generate g_i signals for the partial carry look-ahead unit and the function f_i. The other subcircuit implements the carry-out bit c_i.
  Single stuck-at faults at the inputs/output of the gates of the first subcircuit can result in an erroneous propagate p_i signal, in an erroneous generate g_i signal and in an erroneous value of the function f_i. In accordance with the propagate and generate signals of the next adder cells of the group which contains the adder cell A_i, an error of the propagate/generate p_i/g_i signal of the adder cell A_i will or will not result in an erroneous carry-out signal, which is implemented by the partial carry look-ahead unit. This erroneous carry-out signal is compared with the correct carry-out signal of the last adder cell of the group. The self-checking two-rail checker *TRC* detects the difference and indicates an error ($r_1 = r_2$). An erroneous value of the function f_i always results in an erroneous sum bit s_i and the parity p_s becomes also erroneous. This will be detected by the comparison of p_s with $p_a \oplus p_b \oplus p_c$.

Single stuck-at faults at the inputs/output of the gates of the second subcircuit result in an erroneous carry-out bit c_i. It changes also the sum bit s_i and is the carry-in signal for the next adder cell A_{i+1}. Since the adder cells are designed as the carry-dependent sum adder cells either only the sum bit s_{i+1} is erroneous or both the bits s_{i+1} and c_{i+1} are erroneous. If c_{i+1} is erroneous, also s_{i+2}, or both s_{i+2} and c_{i+2} are erroneous. This may be continued until the carry-out signal of the last adder cell of a group is erroneous. An erroneous carry-out signal of a group of adder cells will be detected by comparison with the corresponding carry signal generated by the partial carry look-ahead unit. If the erroneous carry does not propagate to an erroneous carry-out signal of the last adder cell of a group of adder cells, the number of erroneous bits is odd and it will be detected by parity checking.
A single stuck-at fault at the inputs/output of the XOR-gate for the implementation of the sum bit s_i changes the sum bit s_i. It will also be detected by the comparison of p_s with $p_a \oplus p_b \oplus p_c$.

- All single stuck-at faults of the XOR-tree for the implementation of p_s and all single stuck-at faults of the XOR-tree for the implementation of p_c and of two XOR-gates for the implementation of $p_a \oplus p_b \oplus p_c$:
These faults are detected by the comparison of p_s with $p_a \oplus p_b \oplus p_c$.
- All single stuck-at faults in the two-rail checker *TRC*:
The two-rail checker according to [27] is self-checking.
- Odd errors in the register *S*:
The register *S* is parity-protected and odd errors are detected.

Compared to a carry look-ahead adder without error detection the additional hardware for a parity-checked self-checking carry look-ahead adder with carry-dependent sum adder cells consists of:

- two XOR-trees for determination the input parities p_a and p_b of the operands a and b if the inputs are not yet parity-encoded,
- the adder cells have to be designed as carry-dependent sum adder cells,
- a self-checking two-rail checker to compare the carries implemented by the partial carry look-ahead unit with the carries implemented by the last adder cells of the groups,
- an XOR-tree for the implementation of the parity p_s of the n sum bits,
- two XOR-gates for the implementation of the XOR-sum of the paritites p_a, p_b and p_c.

The necessary area for the look-ahead unit is reduced since instead of a complete look-ahead unit only a partial look-ahead unit has to be implemented.

The number of adder cells in the groups influences the necessary area overhead. For smaller numbers of adder cells in the groups the area is larger and for larger numbers of adder cells the area required is smaller.

Since the adder cells within the groups are connected as ripple adders, the calculation of the sum s in the self-checking carry look-ahead adder with carry-dependent sum adder cells is delayed compared to a carry look-ahead adder without error detection.

Self-Checking Sum Bit-Duplicated Carry Look-Ahead Adder

In a sum bit-duplicated carry look-ahead adder according to [38] the sum bits of the adder cells are (inverted) duplicated and stored in duplicated output registers.

The main advantage of sum bit-duplicated adders is that soft errors directly induced in the output registers by α-particles are detected.

The general structure of self-checking sum bit-duplicated carry look-ahead adder according to [38] is shown in Fig. 4.15.

The input operands $a = (a_0, \dots, a_{n-1})$ and $b = (b_0, \dots, b_{n-1})$ are supposed to be parity-encoded with the corresponding parity bits p_a and p_b. The parity bits p_a and p_b are added modulo 2 to form $p_a \oplus p_b$ of both the input operands. The carry-in signal of the adder is assumed to be duplicated in c_{in}^1 and c_{in}^2.

The sum bit-duplicated self-checking carry look-ahead adder also computes the sum $s = (s_0, \dots, s_{n-1})$ as the inverted sum $\overline{s} = (\overline{s}_0, \dots, \overline{s}_{n-1})$. The sum s and the inverted sum $\overline{s}$ are stored in the corresponding output registers S and $\overline{S}$.

The adder consists of the carry look-ahead unit and n adder cells A_i, $i = 0, \dots, n-1$.

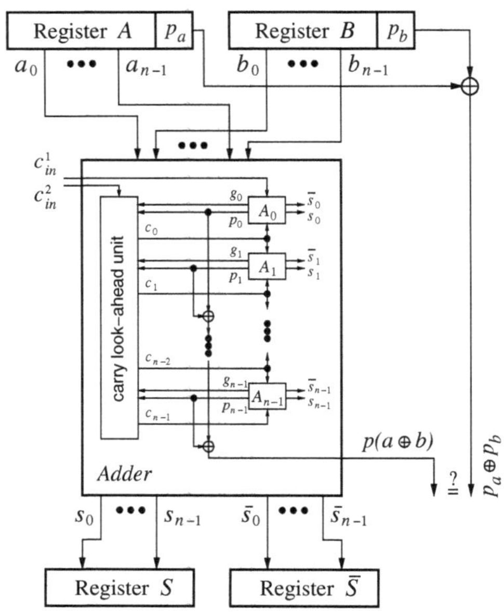

Fig. 4.15 Carry look-ahead adder with sum bit-duplication

The carry look-ahead unit determines the carries $c_0, \ldots, c_{n-1}$ from the duplicated input carry-in signal c_{in}^2 and the propagate p_i and generate g_i signals of the adder cells.

In a sum bit-duplicated carry look-ahead adder for $i = 0, \ldots, n-2$ every adder cell A_i has the two carry-in inputs c_i and c_{i-1}. The carry signal c_i is applied to both the adder cells A_i and A_{i+1}.

The adder cell A_i implements the sum bit s_i, the inverted sum bit $\overline{s}_i$, and for the carry look-ahead unit the propagate signal p_i and the generate signal g_i.

The propagate signals of all adder cells are XORed:

$$
\begin{aligned}
p_0 \oplus p_1 \oplus \ldots \oplus p_{n-1} &= (a_0 \oplus b_0) \oplus (a_1 \oplus b_1) \oplus \ldots \oplus (a_{n-1} \oplus b_{n-1}) = \\
&= (a_0 \oplus a_1 \oplus \ldots \oplus a_{n-1}) \oplus (b_0 \oplus b_1 \oplus \ldots \oplus b_{n-1}) = \\
&= p(a \oplus b) \qquad (4.19)
\end{aligned}
$$

According to (4.19) the XOR-sum $p_0 \oplus p_1 \oplus \ldots \oplus p_{n-1}$ of the propagate signals is equal to the parity $p(a \oplus b)$ of the input operands a and b.

The XOR-sum $p_a \oplus p_b$ determined from the parities p_a and p_b of the input operands and the parity $p(a \oplus b)$ of the input operands determined as the XOR-sum $p_0 \oplus p_1 \oplus \ldots \oplus p_{n-1}$ of the propagate signals are compared.

Every odd error in the input operands will be detected by this comparison and the proposed sum bit-duplicated carry look-ahead adder is code-disjoint.

Also, every odd error of the propagate signals will be detected by this comparison.

The adder cell A_i is shown in Fig. 4.16. The inputs of the adder cell are the corresponding bits a_i and b_i of the operands, and, as already pointed out, the two carry-in signals c_{i-1} and c_i. The propagate signal p_i and the generate signal g_i are implemented by a simple XOR and AND-gate respectively. The sum bit s_i is determined from the propagate signal p_i and the carry-in signal c_{i-1}. To determine the inverted sum bit $\overline{s}_i$, both the carry-in signals c_{i-1} and c_i and the propagate signal p_i are used. As can be seen from Fig. 4.16 the only signal which is shared by different outputs of the adder cell A_i is the propagate signal p_i.

The carry signal c_{n-2} generated by the carry look-ahead unit is the carry-out signal of the considered adder.

The described sum bit-duplicated carry look-ahead adder detects the following errors:

- Odd errors of the input operands:
  Odd errors of the input operands of the adder are detected by the comparison of the parities $p_a \oplus p_b$ and $p(a \oplus b)$.
- All stuck-at faults in the carry look-ahead:
  These faults change the carry signals $c_0, \ldots, c_{n-1}$. Let i_{min} be the minimum index for which the carry is erroneous. Then $c_{i_{min}-1}$ is correct and $c_{i_{min}}$ is erroneous. In the adder cell $A_{i_{min}}$ for the implementation of the sum bit $s_{i_{min}}$ ($s_{i_{min}} = p_{i_{min}} \oplus c_{i_{min}-1}$) the correct carry $c_{i_{min}-1}$ is used. Therefore, the sum bit $s_{i_{min}}$ remains correct. In the same adder cell $A_{i_{min}}$ the inverted sum bit $\overline{s}_{i_{min}}$ is

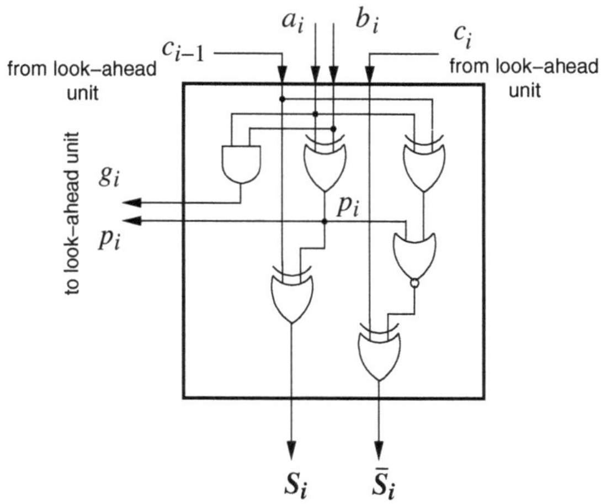

Fig. 4.16 Implementation of the adder cell A_i for the carry look-ahead adder with sum bit-duplication in Fig. 4.15

determined as $\bar{s}_{i_{min}} = (\overline{p_{i_{min}} \vee (a_{i_{min}} \oplus c_{i_{min}-1})} \oplus c_{i_{min}}$ by use of the erroneous carry $c_{i_{min}}$, and the inverted sum bit $\bar{s}_{i_{min}}$ will be erroneous. By the comparing the sum bit $s_{i_{min}}$ with the inverted sum bit $\bar{s}_{i_{min}}$ the considered error due to the considered stuck-at faults in the carry look-ahead unit will be detected.

- All single stuck-at faults in the adder cell A_i of Fig. 4.16:
  The generate signal g_i is implemented in the adder cell by an AND-gate. Single stuck-at faults at the inputs/output of this AND-gate change the generate signal g_i to an erroneous value, and as a result erroneous carries will be produced by the carry look-ahead unit. These errors are detected by comparing the sum bits with the corresponding inverted sum bits as described previously.
  Single stuck-at faults at the inputs/output of the XOR-gate for the implementation of the propagate signal p_i are detected by comparing the parity $p(a \oplus b)$ with the parity $p_a \oplus p_b$. (The parity $p(a \oplus b)$ will be erroneous due to these faults). Thus the propagate signal p_i is always checked, and it can be shared as an input signal for the carry look-ahead unit, for the generation of the sum bit s_i and for the implementation of the inverted sum bit $\bar{s}_i$.
  Single stuck-at faults at the inputs/output of the XOR-gate for the implementation of the sum bit s_i result in an erroneous sum bit s_i. This will be detected by comparing the sum bit s_i with the inverted sum bit $\bar{s}_i$. Single stuck-at faults at the inputs/outputs of the remaining two XOR-gates and the NOR-gate for the implementation of the inverted sum bit $\bar{s}_i$ may erroneously change only the inverted sum bit $\bar{s}_i$. Again, the errors due to this faults are detected by comparing the sum bit s_i with the inverted sum bit $\bar{s}_i$.
  Due to the structure of the adder cell A_i the sum bit s_i and the inverted sum bit $\bar{s}_i$ can never be simultaneously erroneous due to a single stuck-at fault as long as p_i is correct. But if p_i is erroneous, this error will be detected as described.

- All single stuck-at faults of the XOR-tree for the implementation of $p(a \oplus b)$ and of the XOR-gate for the implementation of $p_a \oplus p_b$:
  These faults are detected by the comparison of $p(a \oplus b)$ with $p_a \oplus p_b$.
- All (odd and even) errors in the register S or in the register $\overline{S}$:
  The carry look-ahead adder with sum bit-duplication computes the sum s (stored in the register S) and the inverted sum $\overline{s}$ (stored in the register $\overline{S}$). By the comparison of the contents of the registers S and $\overline{S}$ (in the system) the considered errors will be detected.
  The content of the registers can be compared by additional hardware or at software level.

We emphasize again that all soft errors directly induced in the registers, for instance, by α-particles will be detected by the sum bit-duplicated adder.

The necessary additional hardware for the design of a sum bit-duplicated self-checking carry look-ahead adder compared to a carry look-ahead adder without error detection consists of:

- two XOR-trees for determination the input parities p_a and p_b of the operands a and b if the inputs are not yet parity-encoded,
- additional gates for the implementation of the inverted sum bits of the adder cells,
- an additional XOR-tree for the implementation of the parity $p_0 \oplus p_1 \oplus \ldots\ldots p_{n-1}$ of the propagate signals,
- an additional XOR-gate to determine the XOR-sum $p_a \oplus p_b$ of the input parities p_a and p_b,
- an additional register $\overline{S}$ to store the inverted sum $\overline{s}$.

Since the fan-out of the propagate signal p_i within the adder cell is three and since the carry signals are shared between two successive adder cells, the computation of the sum s in the self-checking sum bit-duplicated carry look-ahead adder is delayed compared to a carry look-ahead adder without error detection.

This section presented different self-checking designs for carry look-ahead adders.

Parity-checked carry look-ahead adders with duplicated carry signals or with carry-dependent sum adder cells were considered in the first two designs. It was shown that these adders are totally self-checking with respect to all single stuck-at faults and with respect to odd errors in the sum bit stored in the output registers.

As a third design, a self-checking sum bit-duplicated carry look-ahead adder was presented. Instead of duplicating the carry bits in this design, the sum bits of the adder cell are duplicated in inverse form. The duplicated sum bits are stored in duplicated output registers. It was demonstrated that the sum bit-duplicated carry look-ahead adder is totally self-checking with respect to all single stuck-at faults and with respect to all (even or odd) errors in the output register. It was shown that the error detection capability of the sum bit-duplicated carry look-ahead adder is almost the same as for duplication and comparison but with a lower area overhead.

All the considered designs are code-disjoint with respect to parity codes.

4.3.2 Self-Checking Partially Duplicated Carry Skip Adder

This section describes a self-checking carry skip adder with parity-encoded input operands according to [39].

It will be shown how the carry skip adder is partially duplicated. A first carry skip adder is completely implemented with adder blocks of "fast" ripple adders as shown in Fig. 4.3. The second carry skip adder utilizes the propagate signals of the first adder, which are implemented only once. It is explained how these propagate signals are used simultaneously to determine the sum bits of both of the partially duplicated adders, to calculate the skip signals of the adder blocks in both adders and to compute the parity of the input operands. It will be shown how the propagate signals can be checked by comparing the *XOR*-sum of these propagate signals with the *XOR*-sum of the input parity bits of the parity-encoded operands.

A carry skip adder without error detection is shown in Fig. 4.17. The adder cells $A_0, A_1, \ldots$ of the carry skip adder are fast ripple adder cells.

The adder cells of a carry skip adder are divided into groups. The carry-in signal of a group of adder cells can bypass the group if all the propagate signals of the adder cells of that group are equal to 1. Figure 4.17 shows a group of four adder cells A_0, A_1, A_2, A_3 (block $B_1(4)$) and the first adder cell A_4 of the next group.

Two types of fast adder cells are used for even and odd adder cells. The single carry-out signal c_i of a traditional adder cell (see for instances Fig. 4.2) is split into two carry signals $C4_i$, $C3_i$ (for odd adder cells) and $C2_i$, $C1_i$ (for even adder cells).

With $c_i = \overline{C3_{i+1} \wedge C4_{i+1}}$ where i is odd and $c_j = C1_{j+1} \wedge C2_{j+1}$ where j is even the correctness of the adder of Fig. 4.17 can be proved by direct calculation.

If a carry is propagated in this adder step by step through a chain of successive adder cells, the delay, which is accumulated in every step, is a gate delay of a single NAND-gate, which is the smallest possible delay per adder cell.

All the inputs $a_0, b_0, \ldots, a_{n-1}, b_{n-1}$ of the adder cells are available at the beginning of the clock cycle. Since the signals $(\overline{a_i b_i}) \wedge (a_{i+1} \oplus b_{i+1})$ or $(a_j \vee b_j) \wedge (a_{j+1} \oplus b_{j+1})$, which are contributing to the computation of the split carry signals depend only on the inputs of their own adder cells the determination of these signals adds only a constant delay to the time, which is needed to complete the computation of the adder.

The input carry $c_{in} = 1$ ($c_{in} = C2_{in} \wedge C1_{in}$) can bypass the considered block $B_1(4)$ of the "fast" carry skip adder in Fig. 4.17 if all the propagate signals p_0, p_1, p_2 and p_3 of the block are equal to 1. To form the bypass (skip) signal the input carry c_{in} and the propagate signals p_0, p_1, p_2 and p_3 are connected by AND-gates. The bypass signal is ORed with both carry-out signals $C2_3$ and $C1_3$ of the last adder cell A_3 of the block $B1(4)$ to implement the carry-in signals $C2'_3$ and $C1'_3$ of the next block. The input carry $c_{in} = 1$ will result in the carry signal $c_3 = 1$ ($c_3 = C2'_3 \wedge C1'_3$) of the second block as soon as the bypass signal arrives this block. If at least one of the propagate signals of $B1(4)$ is equal to 0, the input carry $c_{in} = 1$ will be blocked.

The delay of four "fast" ripple adder cells (delay of four NAND-gates) is almost equal to the delay of the skip signal of a block. If we adapt the method for the

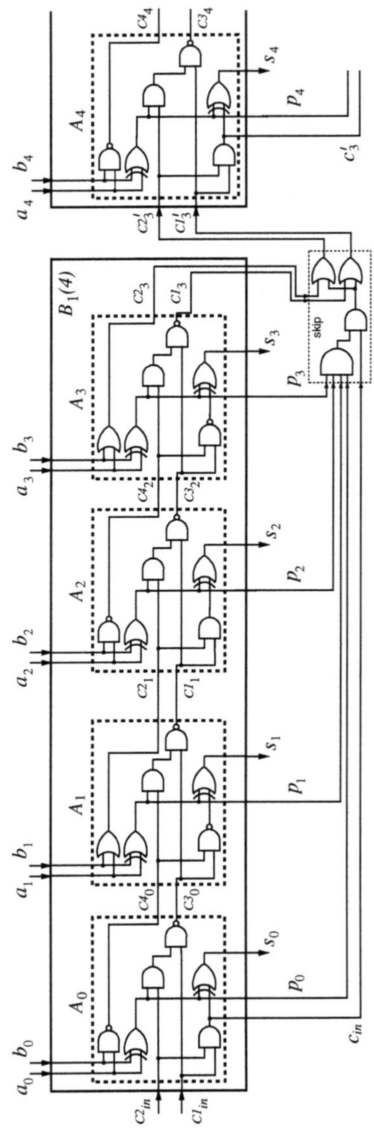

Fig. 4.17 Fast carry skip adder

determination of optimum variable block sizes described in [85] for an n-bit "fast" carry skip adder we can expect the following optimum variable block sizes:

$$k \to (k+4) \to (k+8) \to \ldots \to (k+(t/2-1)4) \to (k+(t/2-1)4) \to \ldots \to (k+8) \to\to (k+4) \to k$$

where $t \approx 2\sqrt{\frac{n}{4}}$ and k is the size of the smallest block.
For $n = 64$ we have to expect 8 blocks of size $2 \to 6 \to 10 \to 14 \to 14 \to 10 \to 6 \to 2$ or 7 blocks of size $4 \to 8 \to 12 \to 16 \to 12 \to 8 \to 4$ as optimum for the "fast" carry skip adder.

The structure of the partially duplicated carry skip adder according to [39] is shown in Fig. 4.18.

A 64-bit adder is shown. The input operands $a = (a_0, \ldots, a_{63})$ and $b = (b_0, \ldots, b_{63})$ are parity-encoded with the corresponding parity bits p_a and p_b. The parity bits p_a and p_b are added modulo 2 to form the input parity $p_a \oplus p_b$. The described self-checking carry skip adder consists of two carry skip adders, a first complete "fast" carry skip adder and a second partially duplicated carry skip adder. Both these adders implement for the same input operands the corresponding sums $s^1 = (s^1_0, \ldots, s^1_{63})$ and $s^2 = (s^2_0, \ldots, s^2_{63})$.

The sums s^1 and s^2 are stored in the output registers S^1 and S^2 respectively. The blocks, which belong to the first "fast" carry skip adder, are denoted by $B^1_j(block\ size)$, where j is the index of a block. $B^2_j(block\ size)$ are the blocks of the second "fast" carry skip adder. The block sizes of both these "fast" carry skip adders are $4 \to 8 \to 12 \to 16 \to 12 \to 8 \to 4$.

The adder cells of the blocks $B^1_j(block\ size)$ of the first complete carry skip adder and of the blocks $B^2_j(block\ size)$ of the partially duplicated carry skip adder are slightly different. The propagate signals $p_i = a_i \oplus b_i$, $i = 0, \ldots, 63$, are only implemented in the adder cells of the first complete carry skip adder but not in the adder cells of the second partially duplicated "fast" carry skip adder.

The propagate signals generated in the blocks $B^1_j(blocksize)$ of the first carry skip adder are used:

- for computing the sum bits s^1_i ($s^1_i = p_i \oplus c_{i-1}$) of the first carry skip adder,
- for computing the sum bits s^2_i ($s^2_i = p_i \oplus c^2_{i-1}$) in the adder cells of the second "fast" carry skip adder,
- for the determination of the skip signals of the blocks $B^2_j(blocksize)$ of the second "fast" carry skip adder,
- for computing the input parity $p(a \oplus b)$ by an XOR-tree of the propagate signals $p_0, \ldots, p_{63}$.

Since the propagate signals $p_0, \ldots, p_{63}$ are only implemented in the adder cells of the first carry skip adder, 64 XOR-gates can be saved. The input parity $p(a \oplus b)$ is implemented by an XOR-tree of the 65 propagate signals $p_0, \ldots, p_{63}$ and not by an XOR-tree of the 128 components $a_0, \ldots, a_{63}$, $b_0, \ldots, b_{63}$ of the input operands a and b. Therefore, another 64 XOR-gates are saved.

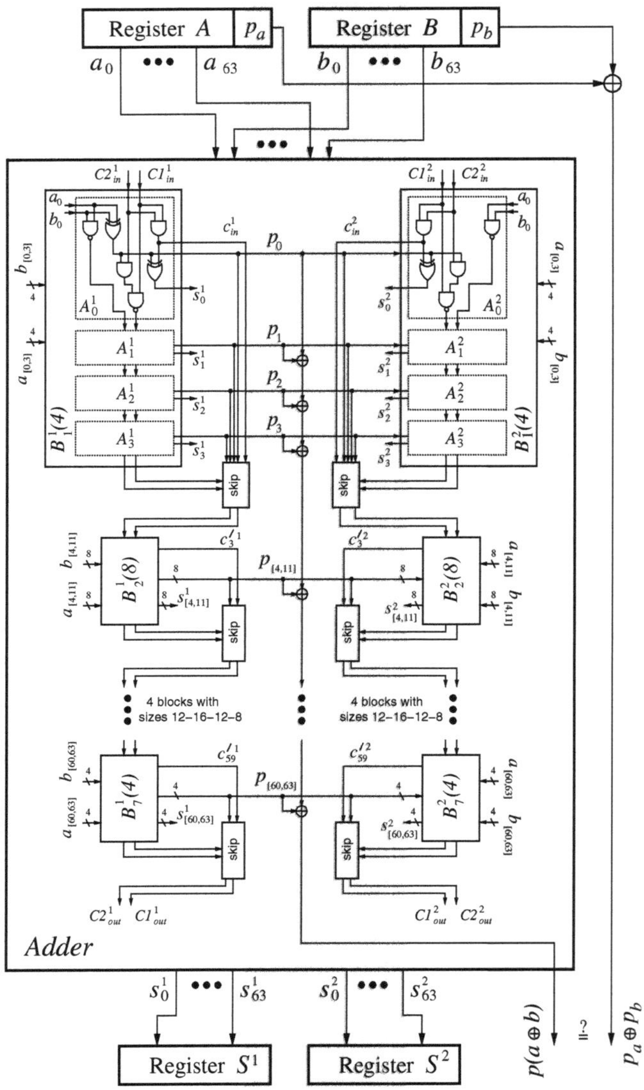

Fig. 4.18 Partially duplicated carry skip adder (for $n = 64$)

The partially duplicated carry skip adder detects the following errors:

- Odd errors of the input operands:
  Odd errors of the input operands of the adder are detected by comparing the parities $p_a \oplus p_b$ and $p_0 \oplus p_1 \oplus \ldots \oplus p_{n-1} = p(a \oplus b)$.
- All single stuck-at faults of the XOR-gates in the adder cells A_i^1, $i = 0, \ldots, 63$, of the first "fast" carry skip adder (the blocks $B_j^1(blocksize)$), which implement the propagate signals $p_0, \ldots, p_{63}$:
  The faults are detected by comparing the parity $p_a \oplus p_b$ with the parity $p(a \oplus b)$.

- All the other single stuck-at faults in the adder cells A_i^1, $i = 0, \ldots, 63$ of the first "fast" carry skip adder:
  Due to these faults the sum s^1 determined by the first "fast" carry skip adder becomes erroneous and the sum s^2 determined by the second partially duplicated carry skip adder remains correct. These faults are detected by comparing the contents of the registers S^1 and S^2.
- All single stuck-at faults in the adder cells A_i^2, $i = 0, \ldots, 63$ of the second "fast" carry skip adder:
  These faults result in an erroneous sum s^2 of the second "fast" carry skip adder, which is detected by comparing the contents of the registers S^1 and S^2.
- Single stuck-at faults in the skip logic "skip":
  The skip logic is implemented twice in the first and in the second carry skip adder. Single stuck-at-1 faults will (for the corresponding inputs) result in an erroneous sum in one of the adders. They are detected by comparing the sums s^1 and s^2 as the contents of the duplicated output registers. A single stuck-at-0 fault of a skip signal is functionally redundant. The correct skip signal which is also generated by the corresponding adder block of the faulty adder and *OR*ed with the value 0 of the erroneous skip signal remains correct. It will be detected if the timing conditions for the computation of the sum bits are violated in the most significant sum bit in the faulty carry skip adder and if the correct value determined by the fault-free adder is compared with the previous values of the sum in the faulty carry skip adder.
- All single stuck-at faults of the XOR-tree for the implementation of $p(a \oplus b)$ and of the XOR-gate for the implementation of $p_a \oplus p_b$ are detected
  These faults are detected by comparing $p(a \oplus b)$ with $p_a \oplus p_b$.
- All single stuck-at faults of the XOR-tree for the implementation of $p(a \oplus b)$ and of the XOR-gate for the implementation of $p_a \oplus p_b$ are detected
  These faults are detected by comparing $p(a \oplus b)$ with $p_a \oplus p_b$.
- All (odd and even) errors in the register S^1 or in the register S^2 are detected
  These faults are detected by comparing the contents of the registers S^1 and S^2 in the system by hardware or software.

The necessary additional hardware for the design of a partially duplicated self-checking carry carry skip adder compared to a carry skip adder without error detection consists of:

- two XOR-trees for determination the input parities p_a and p_b of the operands a and b if the inputs are not yet parity-encoded,
- the "fast" carry skip adder has to be (partially) duplicated, saving n XOR-gates for the implementation of the n propagate signals $p_0, \ldots, p_{n-1}$,
- an additional XOR-tree for the implementation of the parity $p(a \oplus b) = p_0 \oplus \ldots \oplus p_{n-1}$ of the propagate signals,
- one XOR-gate for the implementation of the XOR-sum $p_a \oplus p_b$,
- an additional n bit register S^2 to store the sum s^2 of the partially duplicated "fast" carry skip adder.

Because of the fan-out 4 of the propagate signals p_i, $i = 0, \ldots, 63$ the computation of the sums s^1 and s^2 in the described self-checking carry skip adder is delayed when compared to a carry skip adder without error detection.

This section described a self-checking carry skip adder with parity-encoded input operands.

The carry skip adder was partially duplicated. It it was demonstrated that the propagate signals were used simultaneously to determine the sum bits of both the partially duplicated adders, to calculate the skip signals of the adder blocks in both adders and to compute the parity of the input operands and that the propagate signals can be checked by comparing the *XOR*-sum of the propagate signals with the *XOR*-sum of the input parity bits of the parity-encoded operands. Since the sum is computed twice and stored in a duplicated output register, all soft errors (even and odd) in the output registers are detected. It was explained that single stuck-at-0 faults of the skip signals are functionally redundant and that they are detected if they cause a timing violation. All the other single stuck-at faults are always detected.

4.3.3 Self-Checking Carry Select Adders

Three different designs for self-checking carry select adders will be presented in this section. It will be shown how the duplicated adder blocks for the carry-in signals 0 and 1 of ordinary carry select adders without concurrent checking can be utilized and modified for the design of self-checking carry select adders. In the first design the adder blocks are checked modulo p. In the second design the adder blocks are modified into sum bit-duplicated adders and in the third one the adder blocks for the carry-in signals 1 are replaced by simpler "Add1-circuits" implemented at transistor level.

- *Modulo p-checked carry select adder*
  The two sums determined by the corresponding duplicated adder blocks of a carry select adder for the carry-in signals 0 and 1 differ arithmetically by 1. If these sums would be directly compared, a full adder for every pair of adder blocks would be needed. In the proposed design these sums will be compared modulo p and it will be shown how these two sums can be efficiently compared modulo p by a special "modulo p check and select box" which also selects the correct values of the corresponding sum bits according to the incoming carry signal. It will be explained that for $p = 3$ a much simpler design as in the general case is possible. There are no restrictions for the design of the adder blocks.
- *Sum bit-duplicated carry select adder*
  In the described sum bit-duplicated carry select adder both the original sum of the operands and the inverted sum of the operands will be determined. The input operands are supposed to be parity-encoded and the adder is code-disjoint with

respect to a parity code. It will be shown how all necessary propagate signals for the adder can be computed by a single propagate generator. It will be explained that the remaining parts of the duplicated adder blocks for the carry-in signals 1 and 0 are all implemented as sum bit-duplicated adder blocks with duplicated carry-out signals and how from these components the totally self-checking carry select adder can be designed. All (even and odd) errors in the output registers will be detected additionally to all single stuck-at faults and odd input errors .

- *Sum bit-duplicated carry select adder by use of Add1-circuits*
  In this section an enhancement modification of the sum bit-duplicated carry select adder of the previous section will be presented. It will be shown how the duplicated adder blocks for the carry-in signal 1 can be replaced by much simpler Add1-circuits. The reason for this is that in a carry select adder the arithmetic value of the sum bits generated by an adder block with a carry-in signal 0 differ from the arithmetic value of the sum bits generated by the corresponding duplicated adder block with the input signal 1 arithmetically by 1.
  A possible implementation of an Add1-circuit by use of multiplexors and a special transistor logic block *TLB* will be given.

Modulo p-Checked Carry Select Adder

The structure of a 64-bit *modulo p*-checked carry select adder according to [96, 97] is shown Fig. 4.19.

In a carry select adder the resulting sum bits are, as already described, selected from the groups of sum bits, which are computed in duplicated adder blocks for both the carry-in signals “0” and “1”. These duplicated adder blocks, which already exist in a carry select adder without error detection, are utilized for *modulo p* checking. No restrictions are imposed on the structure of the adder blocks.

In Fig. 4.19 a *modulo p*-checked 64-bit carry select adder is shown.

The duplicated adder blocks $B_2^0(8)$, $B_2^1(8)$; $B_3^0(12)$, $B_3^1(12)$; $B_4^0(12)$, $B_4^1(12)$; $B_5^0(12)$, $B_5^1(12)$ and $B_6^0(12)$, $B_6^1(12)$ are mutually checked by the corresponding “modulo check and select boxes”. The carry signals c_7, c_{15}, $c_{27}, \ldots, c_{51}$ are duplicated in c_71, c_72, $c_{15}1$, $c_{15}2$, $c_{27}1$, $c_{27}2, \ldots, c_{51}1$, $c_{51}2$. Also the first adder block $B_1(8)$ is duplicated in $B_1^d(8)$ and the outputs of the duplicated first adder blocks $B_1(8)$ and $B_1^d(8)$ are compared *modulo p*.

The “modulo check and select boxes” also select in relation to their (duplicated) carry-in signals the sum bits and the carry-out signals of the blocks, and they compute the *modulo p* values of the sum bits and the carry-out signals of both the adders of the blocks. These *modulo p* values are compared by a (self-checking) comparator. The comparison of these *modulo p* signals can be carried out in the next time cycle.

We explain the method of *modulo p* checking for the example of the duplicated adder blocks $B_2^0(8)$ and $B_2^1(8)$.

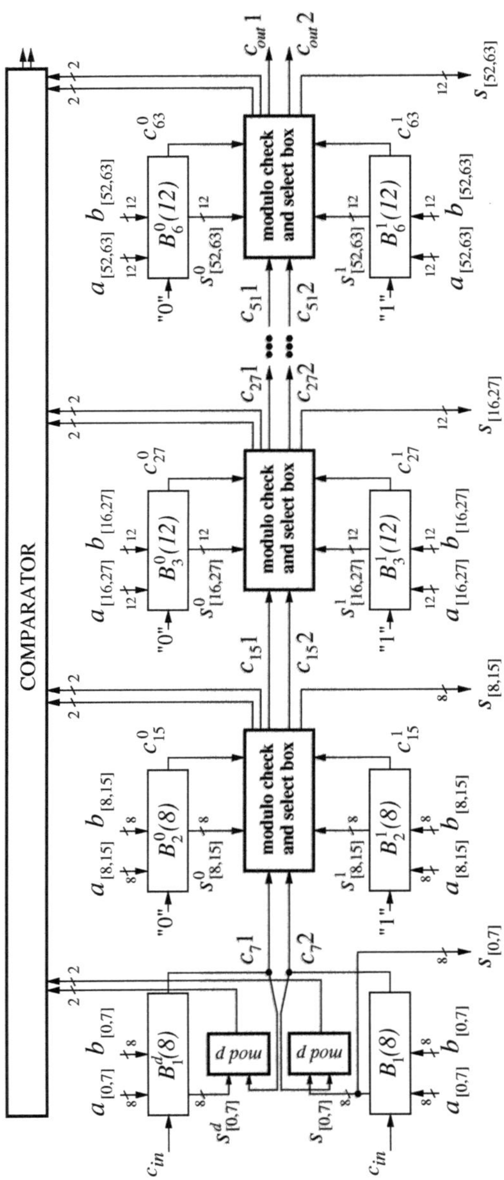

Fig. 4.19 Modulo p-checked carry select adder for $n = 64$ (©IEEE 2003)

The adder blocks $B_2^0(8)$ and $B_2^1(8)$ compute

$$s^0_{[8,15]}, c^0_{15} = a_{[8,15]} + b_{[8,15]}, \tag{4.20}$$

$$s^1_{[8,15]}, c^1_{15} = a_{[8,15]} + b_{[8,15]} + 1, \tag{4.21}$$

and we have

$$s^1_{[8,15]}, c^1_{15} = s^0_{[8,15]}, c^0_{15} + 1. \tag{4.22}$$

The sum and carry bits in the adder blocks $B_2^0(8)$ and $B_2^1(8)$ arithmetically differ by 1. Since the addition of a 1 can affect all the 9 bits of $s^0_{[8,15]}, c^0_{15}$, a direct comparison of $s^1_{[8,15]}, c^1_{15}$ and $s^0_{[8,15]}, c^0_{15} + 1$ is not easy to implement.

Now we explain how these outputs can be compared *modulo p*.

From equation (4.22) we conclude

$$[s^1_{[8,15]}, c^1_{15}] \ mod\ p = [s^0_{[8,15]}, c^0_{15} + 1] \ mod\ p = [s^0_{[8,15]}, c^0_{15}] \ mod\ p \oplus_p 1, \tag{4.23}$$

where $\oplus_p$ denotes the addition *modulo p*. The sum and carry-out bits of the duplicated adder blocks $B_2^0(8)$ and $B_2^1(8)$ differ by 1 *modulo p* only.

Instead of adding arithmetically a 1 to $s^0_{[8,15]}, c^0_{15}$ with a word length of 9 we have to add *modulo p* a 1 to $([s^0_{[8,15]}, c^0_{15}] \ modulo\ p)$ only.

In Fig. 4.20a the "modulo check and select box" is represented. The carry-in signal from the preceding block is duplicated in $c_{in}1$ and $c_{in}2$ respectively, which are the control signals for the multiplexors I, II, II′, III, IV and for the combinational circuits V and V′ for *modulo p* computation. For $c_{in}1 = c_{in}2 = 0$ the circuit V determines $[s^0_{[8,15]}, c^0_{15}] \ mod\ p \oplus_p 1$.

Since the control signal $c_{in}2$ of the block V′ is inverted, this block for $c_{in}2 = 0$ determines $[s^1_{[8,15]}, c^1_{15}] \ mod\ p$. According to equation (4.23) these results are equal as long as no error occurs. Similarly for $c_{in}1 = c_{in}2 = 1$ the circuits V and V′ determine $[s^1_{[8,15]}, c^1_{15}] \ mod\ p$ and $[s^0_{[8,15]}, c^0_{15}] \ mod\ p \oplus_p 1$ respectively. Again, according to equation (4.23) these results are equal as long as no error occurs. The outputs of the combinational circuits V and V′ are to be compared by a self-checking comparator or two-rail checker. The carry-out signal of the blocks is duplicated in $c_{out}1$ and $c_{out}2$. These duplicated carry-out signals are selected by the multiplexors II and II′ with the control signals $c_{in}1$ and $c_{in}2$.

A modification of the "modulo check and select box" is represented in Fig. 4.20b. Only the combinational circuits V and V′ for the *modulo p* computation are modified in the circuits VI and VI′ respectively. For the control signal $c_{in}1 = 0$ the circuits VI and VI′ determine as previously the circuits V and V′ $[s^0_{[8,15]}, c^0_{15}] \ mod\ p \oplus_p 1$ and $[s^1_{[8,15]}, c^1_{15}] \ mod\ p$. For the control signal $c_{in}1 = 1$ the circuits VI and VI′ output $[s^1_{[8,15]}, c^1_{15}] \ modulo\ p \oplus_p (p-1)$ and $[s^0_{[8,15]}, c^0_{15}] \ modulo\ p$. Because of equation (4.23) we have

$$[s^1_{[8,15]}, c^1_{15}] \ modulo\ p \oplus_p (p-1) = [s^0_{[8,15]}, c^0_{15} + 1] \ modulo\ p \oplus_p (p-1) = [s^0_{[8,15]}, c^0_{15}] \ modulo\ p$$

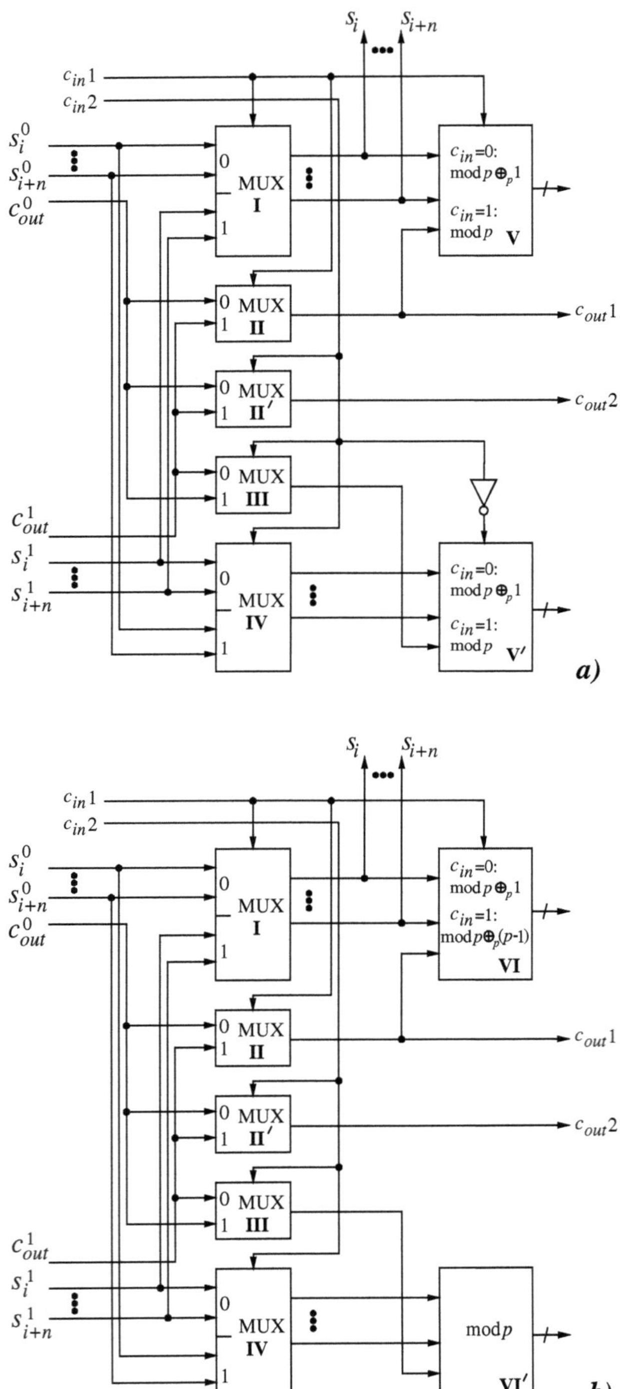

Fig. 4.20 Modulo check and select boxes *a* and *b* (©IEEE 2003)

and the outputs of the circuits VI and VI′ are equal as long as no error occurs. Compared to Fig. 4.20a in Fig. 4.20b no control signal is needed for the circuit VI′.

In the following the case $p = 3$ is considered. For $p = 3$ the outputs of the "modulo check and select boxes" have the word length 2 only, and for a 64-bit self-checking carry select adder a comparator for only two 12-bit words is needed. An advantage for the choice $p = 3$ is also that a special "modulo check and select box" without any additional control line can be used. This special "modulo check and select box" is shown in Fig. 4.21.

We explain how the duplicated blocks $B_2^0(8)$ and $B_2^1(8)$ according to Fig. 4.21 for the carry-in signals $c_{in}1 = c_{in}2 = 0$

$$(\overline{s}_8^0 s_9^0 \ldots s_{15}^0, c_{15}^0) \; mod\; 3 = (1 s_9^1 \ldots s_{15}^1, c_{15}^1) \; mod\; 3 \tag{4.24}$$

and for the carry-in signals $c_{in}1 = c_{in}2 = 1$

$$(\overline{s}_8^1 s_9^1 \ldots s_{15}^1, c_{15}^1) \; mod\; 3 = (0 s_9^0 \ldots s_{15}^0, c_{15}^0) \; mod\; 3 \tag{4.25}$$

are checked.

All the outputs of the carry select adder, either $s_8^0 \ldots s_{15}^0$ for $c_7 1 = c_7 2 = 0$ or $s_8^1 \ldots s_{15}^1$ for $c_7 1 = c_7 2 = 1$ and also the carry-out signals c_{15}^0 and c_{15}^1 are completely checked *modulo 3*.

We show that the equations (4.24) and (4.25) are valid.

First the case $c_{in}1 = c_{in}2 = 0$ is considered.

a. Let $s_8^0 = 0, \overline{s}_8^0 = 1$.

Since

$$s_8^1 s_9^1 \ldots s_{15}^1, c_{15}^1 = s_8^0 s_9^0 \ldots s_{15}^0, c_{15}^0 + 1 \tag{4.26}$$

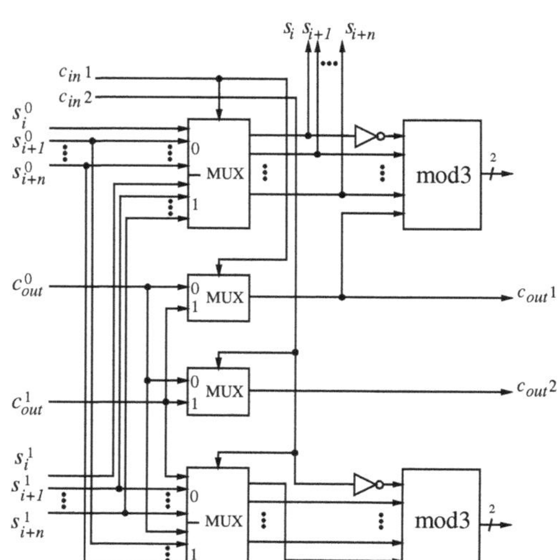

Fig. 4.21 Modulo check and select box *c* for $p = 3$ (©IEEE 2003)

we have $s_8^1 = 1$, $s_9^1 \dots s_{15}^1, c_{15}^1 = s_9^0 \dots s_{15}^0, c_{15}^0$,

$$1s_9^1 \dots s_{15}^1, c_{15}^1 = \bar{s}_8^0 s_9^0 \dots s_{15}^0, c_{15}^0 \tag{4.27}$$

and the equation (4.24) is true.

b. Let now $s_8^0 = 1$, $\bar{s}_8^0 = 0$.

From 4.26 we conclude $s_8^1 = 0$ and we have

$$\begin{aligned}(\bar{s}_8^0 s_9^0 \dots s_{15}^0, c_{15}^0)\ mod\ 3 &= (s_8^0 s_9^0 \dots s_{15}^0, c_{15}^0 - 1)\ mod\ 3 = \\ &= (s_8^0 s_9^0 \dots s_{15}^0, c_{15}^0 + 2)\ mod\ 3 = \\ &= (s_8^1 s_9^1 \dots s_{15}^1, c_{15}^1 + 1)\ mod\ 3 = \\ &= (1 s_9^1 \dots s_{15}^1, c_{15}^1)\ mod\ 3.\end{aligned} \tag{4.28}$$

and the equation (4.24) is also valid in this case.

Now the case $c_{in}1 = c_{in}2 = 1$ is considered.

a. Let $s_8^0 = 0$, $\bar{s}_8^0 = 1$.

From equation (4.26) we have $s_8^1 = 1$ and therefore

$$\begin{aligned}(\bar{s}_8^1 s_9^1 \dots s_{15}^1, c_{15}^1) &= (s_8^1 s_9^1 \dots s_{15}^1, c_{15}^1 - 1) = \\ &= s_8^0 s_9^0 \dots s_{15}^0, c_{15}^0 = 0 s_9^0 \dots s_{15}^0, c_{15}^0\end{aligned} \tag{4.29}$$

and the equation (4.25) is valid.

b. Let now $s_8^0 = 1$, $\bar{s}_8^0 = 0$.

Then we have from equation (4.26) $s_8^1 = 0$ and

$$\begin{aligned}(\bar{s}_8^1 s_9^1 \dots s_{15}^1, c_{15}^1)\ mod\ 3 &= (s_8^1 s_9^1 \dots s_{15}^1, c_{15}^1 + 1)\ mod\ 3 = \\ &= (s_8^0 s_9^0 \dots s_{15}^0, c_{15}^0 + 2)\ mod\ 3 = \\ &= (s_8^0 s_9^0 \dots s_{15}^0, c_{15}^0 - 1)\ mod\ 3 = \\ &= (0 s_9^0 \dots s_{15}^0, c_{15}^0)\ mod\ 3.\end{aligned} \tag{4.30}$$

and the equation (4.25) is true.

The *modulo p*-checked carry select adder detects the following errors:

- All errors due to single stuck-at faults in the "modulo check and select boxes" are immediately detected.
- All errors of the carry-out bits are immediately detected.
- The error detection probability for errors of the sum bits due to single stuck-at faults depends on the value of p and of the concrete implementation of the adder blocks. In a concrete design for p=3 the error error detection probability for errors due to single stuck-at faults in the adder blocks was 99.999% [96, 97].

Modulo p-checked carry select adders are not code-disjoint, and no errors at the inputs of the adder are detected.

Compared to a carry select adder without error detection the area overhead for the implementation of the *modulo p*-checked carry select adder consists of:

- the duplicated first adder block $B_1^d(m)$, where m is the length of the block,
- the "modulo check and select boxes" to perform the modulo checking,
- a comparator to compare the *modulo p* values of the corresponding "modulo check and select boxes".

Advantages of the approach are that no restrictions are imposed on the structure of the adder blocks and that the number of the signals to be compared can be significantly reduced. Thus for $p = 3$ and a 64-bit carry select adder only two 12-bit words are to be compared.

Compared to a carry select adder without error detection the computation of the sum s is not delayed. The comparison of the *modulo p* values can be carried out in the next time cycle.

Sum Bit-Duplicated Carry Select Adder

A self-checking sum bit-duplicated carry select adder was introduced in [98, 40, 99]. In Fig. 4.22 a 64 bit self-checking sum bit-duplicated adder is shown.

The input operands $a = (a_0, \ldots, a_{63})$ and $b = (b_0, \ldots, b_{63})$ are parity-encoded with the corresponding parity bits p_a and p_b. The parity bits p_a and p_b are added modulo 2 to form the parity $p_a \oplus p_b$. The adder also computes for the input operands a and b the sum $s = (s_0, \ldots, s_{63})$ and the inverted sum $\overline{s} = (\overline{s}_0, \ldots, \overline{s}_{63})$, which are stored in the corresponding output registers S and $\overline{S}$. The propagate signals $p_0, \ldots, p_{63}$ of the adder cells are not duplicated. They are implemented from the input operands a and b by the "propagate generator" only once. The XOR-sum of the propagate signals, which is equal to the parity $p(a \oplus b)$ of the operand bits

$$p_0 \oplus p_1 \oplus \ldots \oplus p_{63} = a_0 \oplus b_0 \oplus a_1 \oplus b_1 \oplus \ldots a_{63} \oplus b_{63} = p(a \oplus b),$$

is compared with $p_a \oplus p_b$. Because of this comparison the described sum bit-duplicated adder is code-disjoint, and all odd errors in the input operands are detected.

The corresponding bits of input operands a and b and of the propagate signals of the "propagate generator" are applied to the adder blocks of the sizes of 8, 8, 12, 12, 12 and 12 bits. Unlike the "ordinary" carry select adder, every adder block of the described carry select adder implements the corresponding sum bits and the inverted sum bits. To indicate this all the blocks will be denoted by *SDB*. The carry-out signals of the blocks are duplicated.

The first block $SDB_1(8)$ computes from the operand bits $a_{[0,7]} = a_0, \ldots, a_7$, $b_{[0,7]} = b_0, \ldots, b_7$ and from the propagate signals $p_{[0,7]} = p_0, \ldots, p_7$ the sum bits $s_{[0,7]} = s_0, \ldots, s_7$, the inverted sum bits $\overline{s}_{[0,7]} = \overline{s}_0, \ldots, \overline{s}_7$ and the duplicated carries $c_7 1$ and $c_7 2$ of the block. The carry signal $c_7 1$ of the block $SDB_1(8)$ selects the inverted sum bits $\overline{s}_{[8,15]}$ of the adder and the internal carry signal $c_{15} 1$. These signals are selected from the inverted sum bits $\overline{s}^0_{[8,15]}$ and carry-out signal $c^0_{15} 1$ of the block $SDB^0_2(8)$ and the inverted sum bits $\overline{s}^1_{[8,15]}$ and carry-out signal $c^1_{15} 1$ of the block

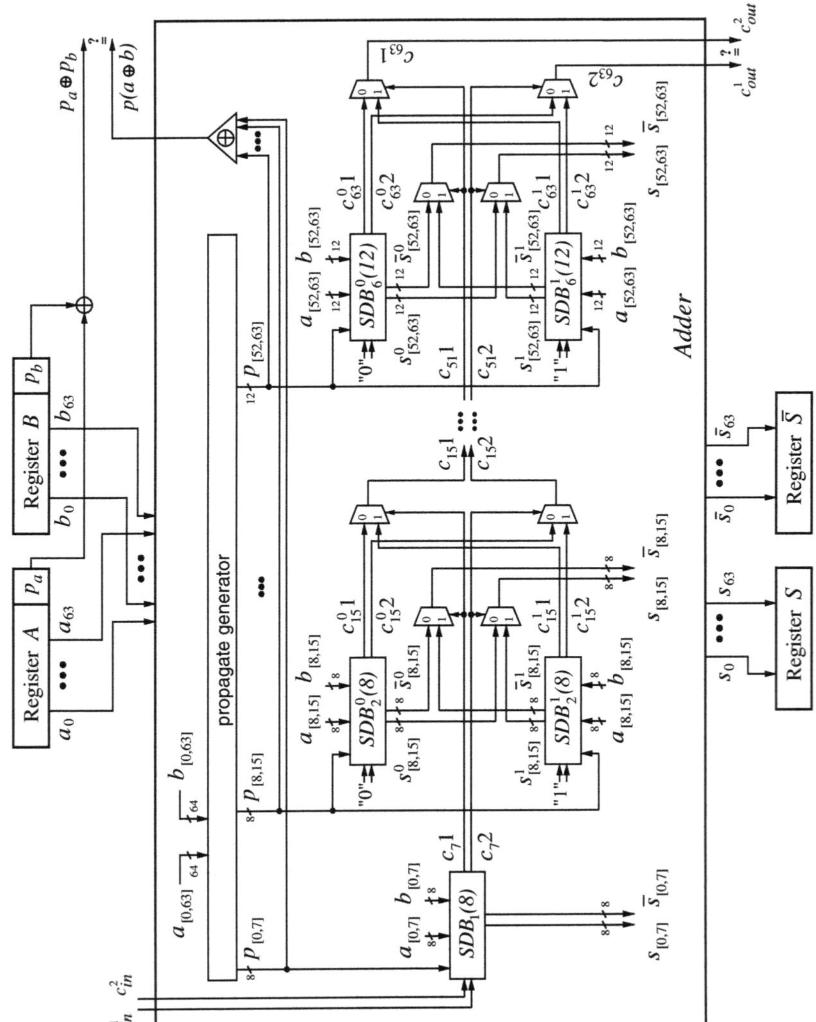

Fig. 4.22 Self-checking sum bit duplicated carry select adder (for $n = 64$)

$SDB_2^1(8)$. The second carry signal c_72 of the block $SDB_1(8)$ selects the sum bits $s_{[8,15]}$ of the adder and the second internal carry signal $c_{15}2$ from the corresponding signals of the blocks $SDB_2^0(8)$ and $SDB_2^1(8)$. The opposite blocks $SDB_2^0(8)$ and $SDB_2^1(8)$ compute the sum bits, the inverted sum bits and the duplicated carry-out signals for carry-in "0" and carry-in "1" respectively. The selection of the remaining sum bits and inverted sum bits of the adder and the duplicated internal carries is performed identically.

The "fast" ripple adder cells of Fig. 4.3 are used for the implementation of the adder blocks. The internal structure of the first adder block $SDB_1(8)$ is shown in Fig. 4.23.

It consists of a first "fast" ripple adder for computing the eight sum bits $s_{[0,7]}$ and the first carry-out signal c_71 and a second "fast" ripple adder with inverted outputs for computing the inverted eight sum bits $\bar{s}_{[0,7]}$ and the duplicated carry-out signal c_72. Both these adders share the propagate signals $p_{[0,7]}$ of the "propagate generator". The propagate signals are not generated internally in the cells. Therefore, one XOR-gate per adder cell can be saved. As carry-in signals for the "fast" ripple adders of the block, the duplicated input carries c_{in}^1 and c_{in}^2 of the carry select adder are used. All the adder blocks $SDB_2^0(8)$, $SDB_2^1(8), \ldots,$ $SDB_6^0(12)$, $SDB_6^1(12)$ are identical to the sum bit-duplicated adder block in Fig. 4.23 with either a constant carry-in signal "0" or "1". For a constant carry-in signal the corresponding sum bit-duplicated adder block can be optimized.

The described sum bit-duplicated carry select adder is totally self-checking with respect to all single stuck-at faults and detects the following errors:

- Odd errors in the input operands:
  If the correct input operands a and b are changed by odd errors into erroneous input operands a' and b' with $p(a \oplus b) \neq p(a' \oplus b')$, then we have $p_a \oplus p_b \neq p(a' \oplus b')$ and the errors are detected.
- All errors due to single stuck-at faults in the "propagate generator":
  These errors change the parity $p(a \oplus b)$ and are detected by the comparison of $p(a \oplus b)$ with $p_a \oplus p_b$.
- All errors due to single stuck-at faults in the adder blocks:
  These errors will be propagated to the sum s or to the inverted sum $\bar{s}$ of the carry select adder and are detected by comparing s and $\bar{s}$.
  If a fault changes only one of the carry-out signals of a block, then this erroneous carry-out signal is an erroneous select signal for the sum selection and it will also be detected by comparing s and $\bar{s}$ in the next adder block.
- All errors due to faults in the multiplexors for the selection of the resulting sum bits and inverted sum bits:
  These faults result in an erroneous selection of the sum bits or the inverted sum bits. The errors due to these faults will again be detected by comparing s and $\bar{s}$.
- All single stuck-at faults in the multiplexors for the selection of the carry-out signals:
  These faults will result in erroneous control signals for the multiplexors for the sum selection. Therefore, they will also be immediately detected by comparing

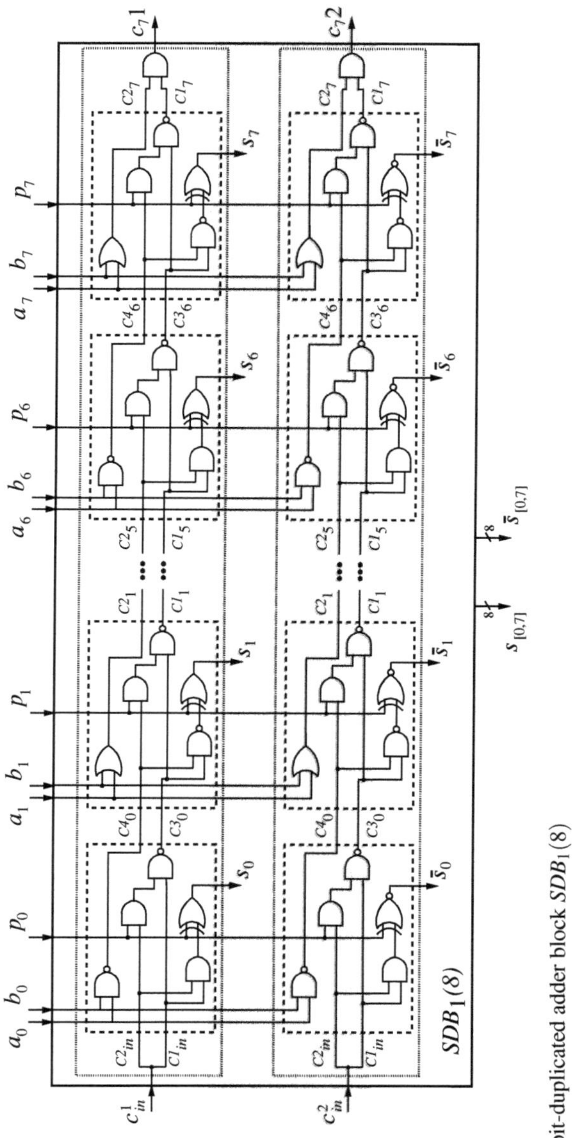

Fig. 4.23 Sum bit-duplicated adder block $SDB_1(8)$

s and $\overline{s}$. The faults in the multiplexors, which select the duplicated output carries of the adder, will be detected by the comparison of c^1_{out} with c^2_{out}.
- All (odd and even) errors in the register S or in the register $\overline{S}$:
  The considered errors will be detected by the comparison of the contents of the registers S and $\overline{S}$.

Special adder blocks are to be used for the design of a self-checking sum bit-duplicated carry select adder. The adder blocks are implemented as duplicated "fast" carry ripple adders and compute the sum bits as well as the inverted sum bits. Compared to a carry select adder without error detection the necessary additional area consists of:

- two XOR-trees for determination the input parities p_a and p_b of the operands a and b if the inputs are not yet parity-encoded,
- one XOR-gate for the implementation of the XOR-sum $p_a \oplus p_b$,
- an additional XOR-tree for the implementation of the parity $p(a \oplus b) = p_0 \oplus \ldots \oplus p_{n-1}$ of the propagate signals $p_0, \ldots, p_{n-1}$,
- a duplicated "fast" ripple adder in each adder block for the implementation of the inverted sum bits and a second carry-out signal,
- additional multiplexors for the selection of the resulting inverted sum bits and of the duplicated carry-out signals,
- an additional register $\overline{S}$ to store the inverted sum $\overline{s}$.

The delay for the computation of the sum bits is not increased compared to a carry select adder without error detection.

Sum Bit-Duplicated Carry Select Adder by Use of Add1-Circuits

The adder blocks of a carry select adder, which calculate the sum bits for the constant carry-in signals "1" can be, as described in [90], replaced by Add1-circuits of Fig. 4.8.

Now we describe how this idea can be utilized for the design of self-checking carry select adders as proposed in [100]. Thereby modified Add1-circuits according to [101] are used.

An efficient design of a modified four-bit Add1-circuit according to [101] is shown in Fig. 4.24. The Add1-circuit is implemented as a simple transistor logic block TLB.

In [101] it is already assumed that the adder blocks with the carry-in signals "0" also compute the sum bits as the inverted sum bits. Therefore, the application of sum bit-duplicated adder blocks fits very well to the design of self-checking carry select adders by use of these modified Add1-circuits. The inputs of the modified Add1-circuit are the sum bits and also the inverted sum bits of the corresponding adder block and the carry-out signal of the preceding adder block. The Add1-circuit now only component-wise selects the appropriate sum bits from the sum bits and

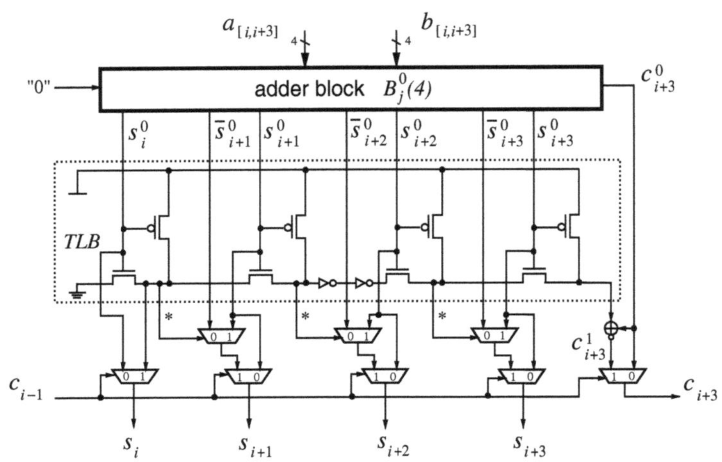

Fig. 4.24 Multiplexor-based Add1-circuit for 4 bits according to [101]

the inverted sum bits that are already generated by the corresponding adder block with the carry-in signal “0”.

A self-checking sum bit-duplicated carry select adder by use of Add1-circuits of Fig. 4.24 is shown in Fig. 4.25.

The input operands $a = (a_0, \ldots, a_{63})$ and $b = (b_0, \ldots, b_{63})$ are again supposed to be parity-encoded with the corresponding parity bits p_a and p_b. The parity bits p_a and p_b are added modulo 2 to form $p_a \oplus p_b$. The adder computes for the input operands a and b the sum $s = (s_0, \ldots, s_{63})$ as well as the inverted sum $\overline{s} = (\overline{s}_0, \ldots, \overline{s}_{63})$, which are stored in the corresponding output registers S and $\overline{S}$.

As in the previous design the propagate signals $p_0, \ldots, p_{63}$ are generated only once. They are determined from the input operands a and b by the “propagate generator”. The XOR-sum of the propagate signals, which is equal to the parity $p(a \oplus b)$ of the operand bits

$$p_0 \oplus p_1 \oplus \ldots \oplus p_{63} = a_0 \oplus b_0 \oplus a_1 \oplus b_1 \oplus \ldots a_{63} \oplus b_{63} = p(a \oplus b),$$

is compared with $p_a \oplus p_b$.

The adder itself consists only of the adder blocks for the computation of the sum under the constant carry-in signals “0”. They are implemented as the sum bit “fast” ripple adders as shown in Fig. 4.23. Each adder block determines the corresponding sum bits and the inverted sum bits. The sum bits and the inverted sum bits of the first adder block $SDB_1(8)$ are the least significant 8 sum bit outputs $s_{[0,7]}$ and the inverted sum bit outputs $\overline{s}_{[0,7]}$ of the adder. The outputs of the sum bit-duplicated adder block $SDB_2^0(8)$ are the inputs of the TLB_2 and TLB_2' circuits. The signals * of the $TLB_2(TLB_2')$-circuit are the control signals for the multiplexors selecting bitwise from $s^0_{[8,15]}$ and $\overline{s}^0_{[8,15]}$ the value $s^1_{[8,15]}(\overline{s}^1_{[8,15]})$. Here the blocks TLB_2 and TLB_2' are identical. Finally the carry-out signal c_72 of the block $SDB_1(8)$ selects the resulting

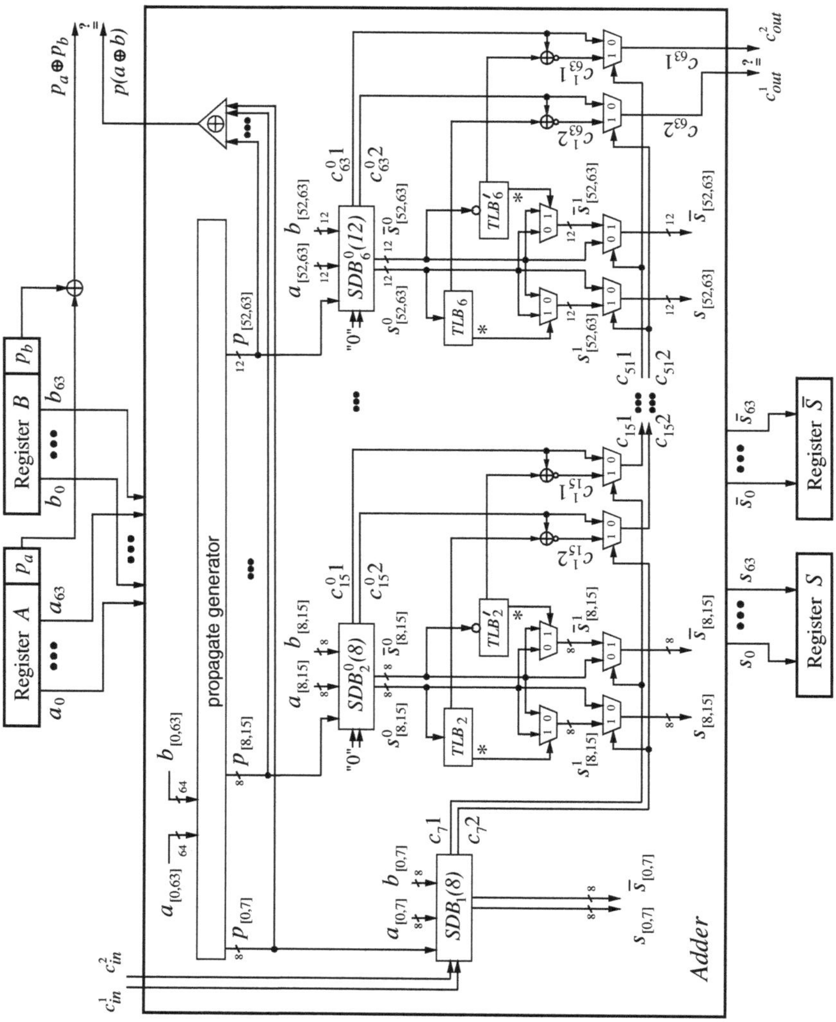

Fig. 4.25 Self-checking sum bit-duplicated carry select adder by use of Add1-circuits (for $n = 64$)

sum $s_{[8,15]}$ by a multiplexor from $s^0_{[8,15]}$ and $s^1_{[8,15]}$, and the carry-out signal $c_{15}2$ from $c^0_{15}2$ and $c^1_{15}2$. The inverted sum bits $\overline{s}_{[8,15]}$ and the carry-out signal $c_{15}1$ are selected by $c_7 1$ from $\overline{s}^0_{[8,15]}$ and $\overline{s}^1_{[8,15]}$, and from $c^0_{15}1$ and $c^1_{15}1$ respectively. In a similar way, the sum bits, the inverted sum bits and the corresponding duplicated carry-out signals of the other blocks are determined. The carry-out signals of the most significant block $SDB^0_6(12)$ are the duplicated output carries of the carry select adder.

The described sum bit-duplicated carry select adder by use of Add1-circuits is totally self-checking with respect to all single stuck-at faults and detects the following errors:

- Odd errors in the input operands:
  In the presence of an odd error the correct input operands a and b are changed into erroneous input operands a' and b' with $p(a \oplus b) \neq p(a' \oplus b')$. Therefore, we have $p_a \oplus p_b \neq p(a' \oplus b')$ and the errors are detected.
- All errors due to single stuck-at faults in the "propagate generator":
  These errors change the parity $p(a \oplus b)$ and are detected by comparing $p(a \oplus b)$ with $p_a \oplus p_b$.
- All errors due to single stuck-at faults in the adder blocks:
  These errors will be propagated to the sum s or to the inverted sum $\overline{s}$ of the carry select adder and detected by comparing s and $\overline{s}$. If a fault changes only one of the carry-out signals of a block, then this erroneous carry-out signal is an erroneous select signal for the sum selection and it will also be detected by comparing s and $\overline{s}$ in the next adder block.
- Errors in TLB-circuits:
  These errors will result in an erroneous selection of a sum bit or inverted sum bit for the case of the carry-in signal "1". Such errors will again be detected by comparing s and $\overline{s}$.
- Errors due to single stuck-at faults in the multiplexors for the selection of the resulting sum bits and inverted sum bits:
  A fault in a multiplexor results in an erroneous selection of a sum bit or an inverted sum bit which will be detected by comparing s and $\overline{s}$.
- Errors due to faults in the multiplexors for the selection of the carry-out signals:
  The carry-out signals are the control signals for the selection of the sum bits or the inverted sum bits. Erroneous control signals are detected by comparing s and $\overline{s}$.
  Errors in the multiplexors selecting the duplicated output carries of the adder will be detected by the comparison of c^1_{out} with c^2_{out}.
- All (odd and even) errors in the register S or in the register $\overline{S}$:
  These errors are detected by comparing the contents of the registers S and $\overline{S}$.

Special adder blocks and Add1-circuits have to be used for the design of the sum bit-duplicated carry select adder. The adder blocks are implemented as duplicated "fast" carry ripple adders and compute the sum bits as well as the inverted sum bits. Compared to a carry select adder without error detection the following additional area is needed:

- two XOR-trees to determine the input parities p_a and p_b of the input operands if the inputs are not yet parity-encoded,
- one XOR-gate for the implementation of $p_a \oplus p_b$,
- an additional XOR-tree for the implementation of the parity $p(a \oplus b) = p_0 \oplus \ldots \oplus p_{n-1}$ of the propagate signals,
- Add1-circuits for the implementation of the sum bits and inverted sum bits for the carry-in signal "1",
- additional multiplexors for the selection of the resulting inverted sum bits and the selection of the duplicated carry-out signals,
- an additional register $\overline{S}$ to store the inverted sum $\overline{s}$

The additional delay caused by the Add1-circuits is almost negligible.

This section presented three different designs for self-checking carry select adders. It was shown how the duplicated adder blocks for the carry-in signals 0 and 1 of ordinary carry-select adders without concurrent checking can be utilized and modified in different ways for the design of self-checking carry select adders. This was demonstrated for a *modulo p-checked carry select adder, a sum bit-duplicated carry select adder* and a *a sum bit-duplicated carry select adder by use of Add1-circuits*. In the first design for the modulo p-checked carry select adder the duplicated adder blocks were checked modulo p. In the sum bit-duplicated carry select adder all the adder blocks are implemented as sum bit-duplicated adders. It was described how a single propagate generator can generate all the necessary propagate signal for the duplicated adder blocks and for checking the input parities of the input operands. It was shown that besides all single stuck-at faults also all soft errors (even or odd) in the duplicated output registers are also detected. In the third design it was demonstrated how the duplicated adder blocks for the carry-in signals 1 can be replaced by simple Add1-circuits.

Section 4.3 described the design of self-checking adders. Since adders are very regular structures this chapter has also demonstrated how the methods of concurrent checking can be adapted to regular structures. All types of practically relevant types of adders were considered and new solutions for the self-checking adder design were given. The design of self-checking carry look-ahead adders, self-checking carry skip adders and self-checking carry select adders was presented. It was demonstrated how the general method of code-disjoint partial duplication with parity checking for the non-duplicated part can be efficiently applied to the design of self-checking adders. It was shown how the special properties of the basic adder equations, the already existing functionally redundant parts of the adders (without error detection), which were implemented to improve the speed of the adders, can be efficiently exploited to achieve optimum results with respect to error detection and additional area overhead. A new adder, the sum bit-duplicated adder was used for concurrent checking and especially for error detection of the soft errors directly induced in the output registers. The best possible state-of-the-art self-checking adders for the different types of adders were presented.

References

1. M. Abramovici, A. D. Friedman, and M. A. Breuer, *Digital Systems Testing and Testable Design*. John Wiley & Sons Inc, 1994.
2. M. L. Bushnell and V. D. Agrawal, *Essentials of Testing for Analog, Logic and Memory VLSI Circuits*. Kluwer Academic Publishers, 2000.
3. P. K. Lala, *Digital Circuit Testing and Testability*. San Diego: Academic Press, 1997.
4. D. Pradhan, *Fault-Tolerant Computer System Design*. Upple Saddle River: Prentice Hall, 1996.
5. P. K. Lala, *Self-Checking and Fault-Tolerant Digital Design*. Morgan Kaufmann Publishers, 2001.
6. I. Koren and C. Krishna, *Fault-Tolerant Systems*. Morgan Kaufmann Publishers, 2007.
7. E. S. Sogomonyan, "The Design of Discrete Devices with Diagnosis in the Course of Operation," *Automation and Remote Control*, vol. 31, no. 10, pp. 1854–1860, 1970.
8. D. Marienfeld, E. S. Sogomonyan, V. Ocheretnij, and M. Gössel, "A New Self-Checking Multiplier by Use of a Code-Disjoint Sum Bit-Duplicated Adder," in *Proceedings of 9th IEEE European Test Symposium*, pp. 73–78, 2004.
9. D. Marienfeld, E. S. Sogomonyan, V. Ocheretnij, and M. Gössel, "New Self-Checking Output-Duplicated Booth Multiplier with High Fault Coverage for Soft Errors," in *Proceedings of 14th IEEE Asian Test Symposium*, (Calcutta), pp. 76–81, 2005.
10. D. Marienfeld, E. S. Sogomonyan, V. Ocheretnij, and M. Gössel, "A New Self-Checking and Code-Disjoint Non-Restoring Array Divider," in *Proceedings of 12th IEEE International On-line Test Symposium*, (Como, Italy), pp. 23–28, 2006.
11. H. Vierhaus, W. Meyer, and U. Gläser, "CMOS Bridges and Resistive Transistor Faults: IDDQ versus Delay Effects," in *Proceedings of International Test Conference*, pp. 83–91, 1993.
12. M. Renovell, P. Huc, and Y. Bertrand, "The Concept of Resistance Interval: A New Parametric Model for Realistic Resistive Bridging Fault," in *Proceedings of 13th IEEE VLSI Test Symposium*, pp. 184–189, 1995.
13. C. Lee and D. Walker, "A PPSFP Simulator for Resistive Bridging Faults," in *Proceedings of 18th IEEE VLSI Test Symposium*, pp. 105–110, 2000.
14. I. Polian, P. Engelke, M. Renovell, and B. Becker, "Modelling Feedback Bridging Faults With Non-Zero Resistance," *Journal of Electronic Testing: Theory and Applications*, vol. 1, pp. 57–69, 2005.
15. J. F. Wakerly, *Digital Design*. Upper Saddle River: Prentice Hall, 2001.
16. P. Lidén, P. Dahlgren, R. Johansson, and J. Karlsson, "On Latching Probability of Particle Induced Transients in Combinational Networks," *Fault-Tolerant Computing - FTSC-24, Digest of Papers*, pp. 340–349, 1994.
17. E. S. Sogomonyan, "Design of Built-In Self-Checking Monitoring Circuits for Combinational Devices," *Automation and Remote Control*, vol. 35, no. 2, pp. 280–289, 1974.

18. E. S. Sogomonyan and M. Gössel, "Design of Self-Testing and On-line Fault Detection Combinational Circuits with Weakly Independent Outputs," *Journal of Electronic Testing: Theory and Applications (JETTA)*, vol. 4, pp. 267–281, 1993.
19. V. Saposhnikov, A. Morosov, V. Saposhnikov, and M. Gössel, "A New Design Method for Self-Checking Unidirectional Combinational Circuits," *Journal of Electronic Testing: Theory and Applications (JETTA)*, vol. 12, pp. 41–53, 1998.
20. M. Nicolaidis, I. Jansch, and B. Courtous, "Strongly Code-Disjoint Checkers," in *Symposium on Fault-Tolerant Computing*, 1984.
21. T. Nanya and T. Kawamura, "Error Secure/Propagating Concept and its Application to the Design of Strongly Fault-Secure Processors," *IEEE Transactions on Computers*, vol. 37, no. 1, pp. 14–24, 1988.
22. P. Nigh and A. Gattiker, "Test Method Evaluation Experiments & Data," in *Proceedings of International Test Conference*, pp. 454–463, 2000.
23. I. Polian, J. Hayes, S. Kundu, and B. Becker, "Transient Fault Characterization in Dynamic Noisy Environments," in *Proceedings of International Test Conference*, pp. 1048–1057, 2005.
24. A. Diril, Y. Dhillon, A. Chatterjee, and A. Singh, "Design of Adaptive Nanometer Digital Systems for Effective Control of Soft Error Tolerance," in *Proceedings of 23rd IEEE VLSI Test Symposium*, pp. 298–303, 2005.
25. M. Agarwal, M. Paul, M. Zhang, and S. Mitra, "Circuit Failure Prediction and its Application to Transistor Aging," in *Proceedings of 25th IEEE VLSI Test Symposium*, pp. 277–284, 2007.
26. C. Lisboa, F. Kastensmidt, E. H. Neto, G. Wirth, and L. Casrro, "Using Built-in Sensors to Cope with Long Duration Transient Faults in Future Technologies," in *Proceedings of International Test Conference*, p. paper 24.3, 2007.
27. W. C. Carter and P. R. Schneider, "Design of dynamically checked computers," in *JFIP Congress*, (Edinburg, Scotland), pp. 878–883, 1968.
28. D. A. Anderson, "Design of Self-Checking Digital Networks Using Coding Techniques," Tech. Rep. 527, CSL/University of Illinois, 1971.
29. S. Kundu, E. S. Sogomonyan, M. Gössel, and S. Tarnick, "Self-Checking Comparator with One Periodic Output," *IEEE Transactions on Computers*, vol. C-45, no. 3, pp. 379–380, 1996.
30. C. Metra, M. Favalli, and B. Ricc, "Highly Testable and Compact Single Output Comparator," in *Proceedings of 15th IEEE VLSI Test Symposium*, pp. 210–215, 1997.
31. S. Matakias, Y. Tsiatouhas, T. Haniothakis, and A. Arapoyanni, "Ultra Fast and Low Cost Parallel Two-Rail Code Checker Targeting High Fan-In Applicatons," in *Proceedings of IEEE CS Annual Symposium on Emerging Trends in VLSI Systems Design (ISVLSI 04)*, pp. 203–206, 2004.
32. D. E. Muller and W. S. Bartky, "A Theory of Asycronous Circuits," in *Proceedings of International Symposium on the Theory of Switching*, pp. 204–243, Harward University Press, 1959.
33. E. S. Sogomonyan and M. Gössel, "Schaltung zum Vergleichen von zwei *n*-stelligen binären Datenwörtern," *Deutsche Patentschrift 10 2005 013 883, G06F 7/04*, 20005/2006.
34. M. P. Kusko, B. J. Robbins, T. J. Koprowski, and W. V. Huott, "99% AC coverage using only LBIST and the 16GHz IBM S/390 zSeries 900 Microprocessor," in *Proceedings of International Test Conference*, pp. 586–592, 2001.
35. F. F. Sellers(Jr.), M. Y. Hsiao, and L. W. Bearnson, *Error Detection Logic for Digital Computers*. McGraw-Hill, 1968.
36. D. P. Siewiorek and R. S. Schwarz, *The Theory and Practice of Reliable System Design*. Bedford: Design Press, 1982.
37. E. S. Sogomonyan and M. Gössel, "New Totally Self-Checking Ripple and Carry Look-Ahead Adders," in *Proceedings of 3rd On-Line Testing Workshop*, pp. 36–40, 1997.
38. E. S. Sogomonyan, V. Ocheretnij, and M. Gössel, "A New Code-Disjoint Sum Bit-Duplicated Carry Look-Ahead Adder for Parity Codes," in *Proceedings of 10th Asian Test Symposium*, pp. 57–62, 2001.

39. D. Marienfeld, V. Ocheretnij, M. Gössel, and E. S. Sogomonyan, “Partially Duplicated Code-Disjoint Carry Skip Adder,” in *Proceedings of 7th IEEE International Symposium on DEFECT and FAULT TOLERANCE in VLSI Systems*, pp. 78–86, 2002.
40. E. S. Sogomonyan, D. Marienfeld, V. Ocheretnij, and M. Gössel, “A New Self-Checking Sum Bit-Duplicated Carry Select Adder,” in *Proceedings of Design, Automation and Test in Europe Conference – DATE*, pp. 1360–1361, 2004.
41. H. Hartje, E. S. Sogomonyan, and M. Gössel, “Code-Disjoint Circuits for Parity Codes,” in *Proceedings of 6th Asia Test Symposium*, pp. 100–105, 1997.
42. K. Mohanram and N. A. Touba, “Cost-Effective Approach for Reducing Soft Error Failure Rate in Logic Circuits,” in *Proceedings of International Test Conference*, pp. 893–901, 2003.
43. W. W. Peterson and E. J. Weldon(Jr.), *Error Correcting Codes*. The MIT-Press, 1994.
44. S. Lin and D. J. Costello, *Error Control Coding*. Pearson US Imports & PHIPEs, 2004.
45. H. H. McClellan and C. M. Rader, *Number Theory in Digital Signal Processing*. Prentice Hall, Englewood Cliffs, 1979.
46. M. Gössel and E. S. Sogomonyan, “A Non-linear Split Error Detection Code,” *Fundamenta Informatica*, vol. 83, pp. 109–115, 2008.
47. J. M. Ashjaee and S. M. Reddy, “On Totally Self-checking Checkers for Separable Codes,” *IEEE Transactions on Computers*, vol. 26, no. 8, pp. 737–744, 1977.
48. H. Fujiwara and A. Yanamoto, “Parity-Scan Design to Reduce the Cost of Test Application,” in *Proceedings of International Test Conference*, pp. 283–292, 1992.
49. V. Moshanin, V. Ocheretnij, and A. Dmitriev, “The Impact of Logic Optimization on Concurrent Error Detection,” in *Proceedings of 4th IEEE International On-Line Testing Workshop*, pp. 81–84, 1998.
50. E. V. Slabakov, “Design of Totally Self-checking Combinational Circuits by Use of Residual Codes,” *Automation and Remote Control*, vol. 40, pp. 1333–1340, 1979.
51. M. Gössel and E. S. Sogomonyan, “Self-Parity Circuits for Self-Testing, Concurrent Checking, Fault Detection and Parity-Scan-Design,” in *Proceedings of 11th IEEE VLSI Test Symposium*, pp. 103–111, 1993.
52. N. Touba, “Logic Synthesis for Concurrent Error Detection,” tech. rep., CLS TN 93x/Center for Reliable Computing, Stanford University, 1992.
53. P. Böhlau, “Zero Aliasing Compression Based on Groups of Weakly Independent Outputs in Circuits with High Complexity for Two Fault Models,” *Lecture Notes in Computer Science*, vol. 852, pp. 289–306, 1994.
54. J. Savir, “Shrinking Wide Compressors,” *IEEE Transactions on Computer-Aided Design*, vol. 14, pp. 1379–1387, 1995.
55. K. Chakrabarty and J. P. Hayes, “Test Response Compaction Using Multiplexed Parity Trees,” *IEEE Transactions on Computer-Aided Design*, vol. 15, pp. 1399–1408, 1996.
56. M. Seuring, M. Gössel, and E. S. Sogomonyan, “A Structural Approach for Space Compaction for Concurrent Checking and BIST,” in *Proceedings of 16th IEEE VLSI Test Symposium*, pp. 354–361, 1998.
57. S. M. Reddy, K. K. Saluja, and M. G. Karpovsky, “A Data Compression Technique for Built-In Self-Test,” *IEEE Transactions on Computers*, vol. 37, pp. 1151–1156, 1988.
58. A. Dutta and N. A. Touba, “Synthesis of Non-Intrusive Concurrent Error Detection Using an Even Error Detecting Function,” in *Proceedings of International Test Conference*, p. Paper 40.3, 2005.
59. H. Hartje, E. S. Sogomonyan, and M. Gössel, “Code-Disjoint Circuits for Parity Codes,” in *Proceedings of 6th Asian Test Symposium*, pp. 100–105, 1997.
60. A. Morosov, M. Gössel, and H. Hartje, “Reduced Area Overhead of the Input Parity for Code-Disjoint Circuits,” in *Proceedings of 5th On-Line Testing Workshop*, pp. 227–230, 1999.
61. M. Gössel, V. Saposhnikov, A. Dmitriev, and V. Saposhnikov, “A New Method for Concurrent Checking by Use of a 1-out-of-4 Code,” in *Proceedings of 6th On-Line Testing Workshop*, pp. 147–152, 2000.
62. A. Morosov, V. Saposhnikov, V. Saposhnikov, and M. Gössel, “New Self-Checking Circuits by Use of Berger Codes,” in *Proceedings of 6th On-Line Testing Workshop*, pp. 141–146, 2000.

63. V. Saposhnikov, A. Morosov, V. Saposhnikov, and M. Gössel, "Concurrent Checking by Use of Complementary Circuits for 1-out-of-3 Codes," in *Proceedings of 2nd IEEE International Workshop on Design and Diagnostic of Electronic Circuits and Systems*, pp. 404–407, 2002.
64. A. Morosov, V. Saposhnikov, V. Saposhnikov, and M. Gössel, "Complementary Circuits for On-Line Detection for 1-out-of-3 Codes," in *Proceedings of ARCS - Organic and Pervasive Computing*, pp. 76–83, 2004.
65. V. Saposhnikov, V. Saposhnikov, A. Morosov, and M. Gössel, "Necessary and Sufficient Conditions for the Existence of Self-Checking Circuits by Use of Complementary Circuits," in *Proceedings of 10th International On-line Testing Symposium*, pp. 25–30, 2004.
66. I. Visirew, "Design of Totally Self-Checking Checkers for Constant-Weight Codes," *Automation and Remote Control*, vol. 16, 1982.
67. M. Marouf and A. Friedman, "Efficient Design of Self-Checking Checkers for a Berger Code," *IEEE Transactions on Computer-Aided Design*, vol. 27, pp. 482–490, 1978.
68. V. V. Saposhnikov, V. V. Saposhnikov, A. Morosov, and M. Gössel, "Necessary and Sufficient Conditions for the Existence of Self-Checking Circuits by Use of Complementary Circuits," tech. rep., University of Potsdam, 2004.
69. S. Graf and M. Gössel, *Fehlererkennungsschaltungen*. Berlin: Akademie Verlag, 1987.
70. M. Gössel and S. Graf, *Error Detection Circuits*. London: McGraw-Hill, 1993.
71. S. Ahmikhahaizim, P. Drinedas, and Y. Makris, "Concurrent Error Detection for Combinational and Sequential Logic via Output Compaction," in *Proceedings of International Symposium on Quality Electronic Design*, pp. 319–324, 2004.
72. D. Reynolds and G. Metze, "Fault Detection Capabilities of Alternating Logic," *IEEE Transactions on Computers*, vol. C-27, pp. 1093–1098, February 1978.
73. V. Saposhnikov, V. Saposhnikov, and M. Gössel, *Self-Dual Discrete Devices*. Saint Petersburg: EnergoAtomIzdat, in rus., 2001.
74. V. Moshanin, V. Saposhnikov, V. Saposhnikov, and M. Gössel, "Synthesis of Self-Dual Multi-Output Combinational Circuits for On-line Testing," in *Proceedings of 2nd IEEE International On-Line Testing Workshop*, pp. 107–111, 1996.
75. V. Saposhnikov, A. Dmitriev, M. Gössel, and V. Saposhnikov, "Self-Dual Parity Checking – a New Method for On-Line Testing," in *Proceedings of 14th IEEE VLSI Test Symposium*, pp. 162–168, 1996.
76. V. Saposhnikov, V. Saposhnikov, A. Dmitriev, and M. Gössel, "Self-Dual Duplication for Error Detection," in *Proceedings of 7th IEEE Asian Test Symposium*, pp. 296–300, 1998.
77. A. Morosov, V. Saposhnikov, V. Saposhnikov, and M. Gössel, "Design of Self-Dual Fault-Secure Combinational Circuits," in *Proceedings of 3rd IEEE International On-Line Testing Workshop*, pp. 233–237, 1997.
78. A. Morosov, V. Saposhnikov, V. Saposhnikov, and M. Gössel, "Self-Checking Circuits with Unidirectionally Independent Outputs," *Journal VLSI Design*, vol. 5, no. 4, pp. 333–345, 1998.
79. H. Fujiwara, *Logic Testing and Design for Testabilty*. The MIT-Press, 1985.
80. S. Mitra, M. Zhang, T. Mak, N. Seiert, V. Zia, and K. Kim, "Logic Soft Errors – a Major Barrier to Robust Platform Design," in *Proceedings of International Test Conference*, p. Paper 28.5, 2005.
81. E. S. Sogomonyan, D. Marienfeld, and M. Gössel, "Fehlererkennung mit Fehlerkorrektur für Soft Errors," *GMM/GI/ITG Fachtagung "Zuverlässigkeit und Entwurf", GMMM-Fachbericht 52*, pp. 185–186, 2007.
82. M. Shams, J. Ebergen, and M. Elmasry, "Optimizing CMOS Implementations of the C-Element," in *Proceedings of International Conference on Computer Design (ICCD)*, pp. 700–705, 1997.
83. M. Shams, J. Ebergen, and M. Elmasry, "Modelling and Comparison CMOS Implementations of the C-Element," *IEEE Transactions on VLSI Systems*, vol. 6, pp. 563–567, December 1998.
84. D. A. Patterson and J. L. Hennessy, *Computer Organization and Design*. Morgan Kaufmann Publishers, Inc., 1998.

85. B. Parhami, *Computer Arithmetic. Algorithms and Hardware Designs.* Oxford University Press, 2000.
86. R. H. Katz, *Contemporary Logic Design.* The Benjamin/Cummings Publishing Company, Inc., 1994.
87. I. Koren, *Computer Arithmetic Algorithms.* A.K.Peters, Natick, MA, 2002.
88. M. J. Smith, *Application-Specific Integrated Circuits.* Adison Wesley, Reading, MA, 1997.
89. A. Chandrakasan, W. J. Bowhill, and F. Fox, *Design of High-Performance Microprocessor Circuits.* IEEE Press, 2001.
90. T. Y. Chang and M. J. Hsiao, "Carry-Select Adder Using Single Ripple-Carry Adder," in *IEE Electronic Letters*, vol. 34, No.22, pp. 2101–2103, 1998.
91. F. F. Sellers(Jr.), M. Y. Hsiao, and L. W. Bearnson, *Error Detection Logic for Digital Computers.* McGraw-Hill, 1968.
92. T. R. N. Rao and E. Fujiwara, *Error Control Coding for Computer Systems.* Prentice Hall, New-York, 1989.
93. M. Y. Hsiao and F. F. Sellers, "The Carry-Dependent Sum Adder," *IEEE Transactions on Electronic Computers*, vol. EC-12, pp. 265–268, June 1963.
94. M. Nicolaidis, "Efficient Implementations of Self-Checking Adders and ALU's," in *Symposium on Fault-Tolerant Computing*, pp. 586–595, 1993.
95. V. Ocheretnij, M. Gössel, and E. S. Sogomonyan, "Code-Disjoint Carry-Dependent Sum Adder with Partial Look-Ahead," in *Proceedings of 7th International On-line Testing Workshop*, pp. 147–152, 2001.
96. V. Ocheretnij, M. Gössel, E. S. Sogomonyan, and D. Marienfeld, "A Modulo *p* Checked Self-Checking Carry Select Adder," in *Proceedings of 9th International On-line Testing Symposium*, pp. 25–29, 2003.
97. V. Ocheretnij, M. Gössel, E. S. Sogomonyan, and D. Marienfeld, "Modulo p=3 Checking for a Carry Select Adder," *Journal of Electronic Testing: Theory and Applications*, vol. 22, pp. 101–107, 2006.
98. E. S. Sogomonyan, D. Marienfeld, V. Ocheretnij, and M. Gössel, "A New Self-Checking Sum Bit-Duplicated Carry Select Adder," in *University of Potsdam, Preprint 005/2003, ISSN 0946-7580*, 2003.
99. E. S. Sogomonyan, D. Marienfeld, V. Ocheretnij, and M. Gössel, "Self-Checking Carry Select Adder with Sum Bit-Duplication," in *Proceedings of ARCS - Organic and Pervasive Computing*, pp. 84–91, 2004.
100. V. Ocheretnij, D. Marienfeld, E. S. Sogomonyan, and M. Gössel, "Self-Checking Code-Disjoint Carry Select Adder with Low Area Overhead by Use of Add1-Circuits," in *Proceedings of 10th International On-line Testing Symposium*, pp. 31–36, 2004.
101. Y. Kim and L. S. Kim, "A Low Power Carry Select Adder with Reduced Area," in *Proceedings of International Symposium on Circuits and Systems (ISCAS)*, pp. 218–221, 2001.

Index

α-particles, 31

Add1-circuit, 134, 167
adders, 4
  "fast" ripple adder, 126, 127, 151, 165
  carry look-ahead adder, 4, 126, 127, 138
    carry look-ahead unit, 127, 128
  carry ripple adder, 4, 126
  carry select adder, 4, 126, 132, 156
    modulo p-checked carry select adder, 156, 157
    sum bit-duplicated carry select adder, 156, 163
  carry skip adder, 4, 126, 130
    multi-level skip adder, 132
    skip logic, 130
  carry-dependent sum adder, 137, 138, 143, 146
  carry-duplicated adder, 4
  full adder, 125, 126
  sum bit-duplicated adder, 4, 167
    sum bit-duplicated carry look-ahead adder, 138, 139, 147
alternating inputs, 102, 107
area, 3, 31, 32, 167, 170
  area overhead, 162

C-element, 4, 35, 36, 117, 118
carry duplication, 137
check bits, 56, 57
  duplicated check bits, 119
code-disjoint, 3, 4, 26, 27, 77, 136, 148, 163
  code-disjoint circuits, 6
  code-disjoint partially duplicated circuits, 40
codes, 42
  m-out-of-n code, 49
  1-out-of-3 code, 87
  Berger Code, 46
  block codes, 42
  code checker, 84
  code distance, 43
  code word, 42, 44
  error correction codes, 2
  group parity codes, 3, 45
  linear codes, 44
  non-linear codes, 3, 44
  non-linear split error detection codes, 42, 49
  non-systematic codes, 91
  parity codes, 3, 42, 44
  systematic block codes, 3, 44, 56, 85, 91, 100, 118
comparator, 3, 31–33, 36, 56, 58, 65
  comparator with single dynamic periodic output, 35
  comparators with a single dynamic periodic output, 35
complementary circuits, 3, 84, 85, 90, 100, 102, 107
concurrent checking, 1, 13, 25
  self-testing, 26, 71, 91
controlling value, 5, 7
correction, 4
corrector, 119

D-latch, 117
delay
  maximum delay, 130
distance, 10
duplicated carries, 138, 139
duplication and comparison, 2, 3, 31–33, 41, 46, 100, 150
dynamic output, 36

electrical condition, 22
equality checker, 37
error detection, 40
error detection circuit
  optimal error detection circuit, 98
error detection function, 98
error detection probability, 3, 25, 32
error signal, 32
errors, 3, 43
  error automaton, 24
  error function, 23, 24
  error vector, 43
  even errors, 54, 58, 75, 76, 150, 155, 167, 170
  functional error model, 3, 4, 6, 23
  functional errors, 5
  input errors, 26
    even input errors, 136, 157
    odd input errors, 136, 140, 145, 148, 154, 157, 165, 170
  odd errors, 54, 58, 75, 76, 82, 142, 146, 148, 150, 155, 167, 170
  propagated, 5, 8, 9
  soft errors, 4, 6, 22, 31, 40, 41, 78, 116, 119, 139, 147
  stimulated, 5, 8, 9

fault tolerance, 2
fault-secure, 6, 26
faults, 3, 5, 6
  bridging faults, 3, 5, 6, 10, 13, 23
    bridging between lines, 5
    input-output bridging, 11, 13
    wired AND, 10
    wired OR, 10
  broken lines, 5
  crosstalk, 5, 22
  delay faults, 3, 21, 22
    gate delay faults, 6, 22
    path delay faults, 6, 22
  fault model, 6, 57
  faults at input lines, 107
  faults caused by α-particles, 5
  non-modelled faults, 27
  permanent faults, 1, 5
  stuck-at faults, 3, 109, 141, 145, 148, 154, 155, 165, 170
    single stuck-at faults, 5–7, 12, 27, 112, 142, 145, 146, 149, 150, 155, 157, 162, 170
  transient faults, 1, 5, 22, 116, 119
  transistor faults, 3
    stuck-on faults, 3, 6, 19, 21
    stuck-open faults, 3, 5, 19–21

generalized circuit graph, 3, 54, 60, 61, 71
generate signal, 125
generator, 56, 58, 59, 100, 119
generator circuit, 54, 55
GND, 17
group parity, 24, 53

Hamming distance, 43
Hamming weight, 43

information bits, 57

joint implementation, 72, 79, 103, 108, 109
joint optimization, 109

logic condition, 22

majority voter, 2
modulo p checking, 47

netlist of gates, 56
node splitting, 3, 55, 71
non-controlling value, 7–9

on-line detection, 2
oscillating signal, 5
output dependencies, 3, 6, 24, 62
  functionally independent outputs, 63
  independent outputs, 3, 6, 24, 55, 62
    group of independent outputs, 24, 63
  structurally independent outputs, 63, 64
  unidirectionally independent outputs, 3, 6, 24, 25
  weakly independent outputs, 3, 6, 24, 55, 62, 64, 66
    group of weakly independent outputs, 65, 66, 73
overhead, 31

parity, 53, 65, 72, 75, 78, 82, 120, 135
  input parity, 78
parity bit, 57
parity prediction, 4, 31, 40, 57, 59, 107, 108, 139
parity-checked adder, 136
partial duplication, 3, 4, 31, 40, 78, 139
  partially duplicated circuits, 82, 151, 153
power consumption, 3, 31, 32
predictor, 56, 58, 59, 65, 100, 119
predictor circuit, 54, 55
  parity predictor, 54, 75
propagate signal, 125
  group propagate signal, 129
  propagate generator, 163, 165, 168

self-checking, 27

totally self-checking, 6, 26, 27, 93
self-dual
dual functions, 103, 104
self-dual Boolean functions, 102, 103
self-dual circuits, 102, 104, 109, 111
self-dual complement, 105–108
self-dual duplication, 4, 102, 109
self-dual error detection, 4
self-dual fault-secure circuits, 103, 109, 112, 113
self-dual fault-secure output, 112
self-dual functions, 103
self-dual parity, 4, 102, 107
self-testing, 6, 26
separate implementation, 59, 79, 103, 109
separate optimization, 109
short to ground, 5
short to power supply, 5
side-inputs, 8, 9
single event transition, 6
single event upset, 22
state machine, 11
static output, 36
switches, 18

testing, 1
n-detection testing, 6, 27
"exhaustively" tested, 12
Built-In Self-Test, 1
time redundancy, 102
timing condition, 22
totally self-checking checker, 91–93
transistor
drain, 19
n-net, 17
n-transistors, 17
p-net, 17
p-transistors, 17
source, 19
triple modular redundancy, 2
two-rail, 46
two-rail checker, 32, 33, 37, 142
two-rail logic, 39

undefined output, 18
undefined value, 6

VDD, 17

MIX
Papier aus verantwortungsvollen Quellen
Paper from responsible sources
FSC® C105338

If you have any concerns about our products,
you can contact us on
ProductSafety@springernature.com

In case Publisher is established outside the EU,
the EU authorized representative is:
Springer Nature Customer Service Center GmbH
Europaplatz 3, 69115 Heidelberg, Germany

Printed by Libri Plureos GmbH
in Hamburg, Germany